The Star as Icon

The Star as Icon

CELEBRITY IN THE AGE OF MASS CONSUMPTION

Daniel Herwitz

COLUMBIA UNIVERSITY PRESS   *NEW YORK*

COLUMBIA UNIVERSITY PRESS
Publishers Since 1893
New York   Chichester, West Sussex
Copyright © 2008 Columbia University Press
All rights reserved

Library of Congress Cataloging-in-Publication Data

Herwitz, Daniel Alan, 1955–
    The star as icon: celebrity in the age of mass consumption /
Daniel Herwitz.
        p. cm.
    Includes bibliographical references and index.
    ISBN 978-0-231-14540-4 (cloth : alk. paper)
  1. Fame. 2. Celebrities. 3. Celebrities in mass media.
  4. Aesthetics. 5. Popular culture. I. Title.

BJ1470.5.H47 2008
306.4—dc22   2008012171

Printed in the United States of America
c 10 9 8 7 6 5 4 3 2 1

References to Internet Web Sites (URLs) were accurate at the time of
writing. Neither the author nor Columbia University Press is responsible for Web sites that may have expired or changed since the book
was prepared.

This book is dedicated to Ted Cohen
teacher and friend

Contents

Preface and Acknowledgments

THIS BOOK is about the star icon, Princess Diana, Jackie, Marilyn, Grace Kelly, that endlessly talked about and little understood persona, object of adulation, fantasy, and cult. It begins with a narrative of Lady Diana's funeral, but it is not a "Diana book." It takes off from the pairing of Diana and Grace Kelly, the one a royal who seemed to carry, as if by proxy, the mantle of film star; the other a film star effortlessly become Monaco royal. This transfer or modulation of aesthetic feature from one crown to another is the joint collaboration of media society and public imagination, hence the book's title. Other books have been written about the star icon's celebrity, which we pretty well know how to understand. However, no book has been written that seeks to cut through the gossip, the tabloids, and critical canons of scholarship to focus on her aesthetic formation: what it is about film and television culture, the star system, and consumer society that have made the star icon what she is. Recruiting a philosopher's interest in the media, an ironist's eye on society, and a love of popular culture, my book is an essay on our yearning for and consumption of such iconic figures.

Stories of celebrity and star culture tend to collapse the star icon into a general formula, losing all sense of her uniqueness. Celebrity culture we know how to understand, the star icon we do not. She is a celebrity, but also something quite different, a being, the book argues, caught between transcendence and trauma in her own life and in the public's gaze on her. An effervescent film star living on a distant, exalted planet, she is at the same time a melodrama-soaked soap opera queen whose dismal life she is ever trying to flee or overcome and into the mire of which she constantly sinks—always with the help of the media. The very public that edges her on also secretly desires her to fall apart, since it will be the culmination of a whopping good story. This double life of the icon is sustained by a special alchemy between film and television detailed by the book. In its picture of opposing tensions and strange synergies between film, TV, and consumer society, the book understands aesthetics as a *complex system* and the star icon's emergence as a product (or fault, depending on how you look at it) of the system as a whole. Along the way the reader will, it is hoped, encounter new perspectives on film and the aura of the film star, television, talk show, and serial, and the religious glow of the star in an age of consumer society.

The book is about the way society creates its aesthetic types and also about how it destroys them. It is about grace and it is about cruelty, about the public's longing for charisma (in the guise of religion) and the indifference of its consumer stance. Written by a man in love with popular culture but also deeply critical, it is hoped that the paradoxes and controversies it details will keep the reader thinking even when the silver screen goes grainy, the TV is turned off, and the Warhol painting is left speaking only to the cold blue halogen light of the museum in its off-hours.

A book in multiple registers, multiple persons have aided in its creation. Parts were presented to a Hollywood audience at the house of John Rich, legendary television director and generous friend to the institute I direct, other parts to a philosophical group of aestheticians at the University of Michigan. The fellows of the Institute for the Humanities at Michigan combed over the manuscript in progress. Friends and colleagues gave generously of their time: Gregg Horowitz, Lydia Goehr, Nicholas Delbanco, Michael Steinberg, Ed Dimendberg, Marcia Kinder, Marjorie Perloff, Michael Perlstein, David Gritten, Laurence Goldstein, Kendall Walton, Everett Kramer, and above all, Lucia Saks, inamorata deluxe and companion in channel hopping.

My junior copywriter Sophia Saks-Herwitz helped with the book's title. A part of a chapter on film has appeared in similar guise in my book *Key Concepts in Aesthetics*. Other than that, the material is new.

The Star as Icon

one

The Candle in the Wind

THE FUNERAL interests me most. For it was the culmination of a life known by her admiring public wholly through the media.

Over a million people lined the route to pay homage as her cortege slowly wound its way from Kensington Palace to Westminster Abbey, where the church service began at eleven in the morning. Standing in silence, those crowding the edges of the road bowed their heads when the coffin, draped in a yellow and red standard and topped with white lilies, rolled past. Cameras craned above heads to snap photos of remembrance, tears were shed, prayers

FIGURE 1.1 Princess of Wales's funeral procession passing St. James Park. Photograph by Jialiang Gao, Wikipedia, public domain

whispered. Above the hush could be heard only the clip-clop of the horses and the even rumbling of the carriage. Wheels grinded against stone pavements as horse, soldier, and mourner moved in a single dirge. The horses had been specially trained not to react to the bouquets of flowers thrown from the crowds. The faces of the soldiers were flexed and grim. Her two sons looked well bred and brave. As the cortege passed Buckingham, the royal family stood stone faced outside the palace gate, betraying nothing.

An estimated 2.5 billion people watched on television. When the cortege turned toward the west door of Westminster it stopped, and gun carriage guards approached the coffin to lift and transport it into the church. BBC reporter Tom Fleming began to speak in a well-tempered baritone:

> On a September day in 1982 I described the scene of the first official visit overseas of Diana, Princess of Wales. She was twenty-one and representing the queen at the funeral of a beautiful and much loved princess who had died some days before in a tragic car accident, Princess Grace of Monaco. Little did I imagine that fifteen years later to the month I would be watching the arrival of a simple coffin draped with *another* royal standard, bearing the body of that beautiful and much loved young princess of our own country, killed six days ago, in a tragic car accident.[1]

The guards raised the coffin from the carriage and solemnly transported it to a bier in the center of the abbey. Waiting for the service to begin, the camera again panned inside the church seeking personages. Some had already been shown arriving, including Diana's natural mother—stiff, expressionless, just possibly sober. Now we saw her stepmother Raine, daughter of Barbara Carltand, who wrote over a hundred Romance novels reclining in the satin pillows of her pink couch, dictating chapter and verse to a bevy of secretaries. Margaret Thatcher was there, and Luciano Pavarotti, Donatella Versace, Richard Branson, a wide variety of people who had befriended the princess, those from TV and media who had been both friend and foe, patron and pursuer. In a gelato of language only the BBC could whip up, former secretary of state for health Virginia Bottomley was described as entering the church from "behind." Then it was time for the service. The archbishop read, and one of Diana's sisters, and Tony Blair. An aria from Verdi's *Requiem* was sung. Elton John crooned Bernie Taupin's lyrics, which had been rewritten from their original use in the Versace funeral for Diana's, thus allowing the candle to burn in the wind at both ends of the Atlantic, and Diana's funeral to become the Versace Requiem, "Deus ex machina."

When Elton John finished his ballad a Monaco turn was made in the service and Charles, Earl of Spencer, Diana's younger brother, rose to the podium to deliver the eulogy. The earl had flown up from Cape Town, South Africa, where he sows his wild oats in the same province as Mark Thatcher, son of Margaret. Mark was not at the funeral, perhaps because he was busy toppling some African kingdom with an AK-47 in one hand and his mother's steel reinforced handbag in the other. The earl rose to his finest hour, speaking with passion and precision in a grieving but also celebratory voice, for his eulogy was of a life extinguished yet radiant. Beneath his pitch perfect accent the earl's outrage at Diana's treatment by the royal family was patent. He lambasted—without naming them—those royals who had driven his sister to her eating disorders and depressions. He pledged to protect her "beloved boys," to do everything in his power to ensure they would not "suffer the anguish that used regularly to drive you to tearful despair." He would seek to complete their education, to give them the experiences of life that would allow them to "sing openly, as you'd planned." "We fully respect the heritage into which they have both been born," he said, "but we, like you[, Diana], recognize the need for them to experience as many different aspects of life as possible to arm them spiritually and emotionally for the years ahead."

When Charles turned to reminiscence his voice was the little brother's, forever in love with his older sister:

The last time I saw Diana was on July 1, her birthday in London, when typically she was not taking time to celebrate her special day

with friends but was guest of honour at a special charity fundrais-
ing evening. She sparkled, of course, but I would rather cherish
the days I spent with her in March when she came to visit me and
my children in our home in South Africa. I am proud of the fact
that . . . we managed to contrive to stop the ever-present paparazzi
from getting a single picture of her—that meant a lot to her.

These were days I will always treasure. It was as if we had been
transported back to our childhood when we spent such an enormous
amount of time together—the two youngest in the family. Fundamentally she had not changed at all from the big sister who mothered
me as a baby, fought with me at school, and endured those long train
journeys between our parents' homes with me at weekends.

Intimate childhood bonds were invoked. But elsewhere the earl
spoke of his sister in the language of an adoring fan, speaking to
those millions of millions outside the church or glued to the telly.
Charles' language was starry-eyed, dazed at the mystery of transitions that had been the fabric of her existence, an existence he
called "the most bizarrelike life imaginable after her childhood."
"The world," he said, "cherished her for her vulnerability, while admiring her honesty." It was as if she had created a public of vast
intimates, each overcome by her unapproachable beauty, each simultaneously believing her their intimate. "For such was her extraordinary appeal that the tens of millions [2.5 billion] people taking part in this service all over the world via television and radio,
who never actually met her, feel that they too lost someone close
to them in the early hours of Sunday morning."

Her troubles were of course the stuff of public knowledge and
public sympathy—but not just her troubles, their expression, their
very physiognomy. "Diana explained to me once," he said, "that it
was her innermost feelings of suffering that allowed her to connect
with her constituency of the rejected." Critical to that ability was
her facial register of everything inside, her body posture, as if her
physical incarnation were a mirror of her abundance of feelings,
the tide of vulnerabilities she was wholly unable to contain. In her
face the story already publicly known could be read and reread.

FIGURE 1.2 Flower bouquets for the late Princess Diana. Photograph by
Ralf-Finn Hestoft, Corbis

The nerve endings and musculature of her sculpted cheeks, lips,
and jaw were a living script, not simply a thing of classical beauty.
Only a few actors and actresses have this natural ability, this expressive physiognomy. Anna Magnani had it, and Edith Piaf, and
perhaps Marilyn Monroe. They could croon without makeup, and
one always felt the spontaneity of the genuine in their gestures.
People therefore *believed*: believed without doubting.

The camera has always tracked physiognomy, something Erwin
Panofsky already understood back in 1934.[2] Jimmy Stewart's twitch,
Gary Cooper's tight-lipped jaw, John Wayne's swagger, Meryl
Streep's melting smile: these are what speak cinematic volumes,
bring home and personalize larger narratives. Human physiognomy
is the sculpture of the screen, its visual aria. Her tune was that of a
blonde, Grecian beauty whose thin (underweight) fragility allowed
pain to contort it at the slightest pressure without burying the flame
in the eyes, the candle in the wind. Hers was a face that one felt was
tuned to every register of emotion. Seeing was, in her case, believing, since her body seemed incapable of deliberation, therefore of

deception. She was an actor, yes, played the role of royal on a million occasions, demurred dutifully, held her tea with the right form of expression, sipped and smiled. And yet the façade of royalty, also part of her, never fully erased the body's own language. This was the source of her "integrity," her ability, as in a film actor, to inspire conviction in a willing audience (of billions). Diana's voice was tight, slow, without a great deal of lilt, lacking in wit. She never quite *got* things, shared the silent film comedian's sense of world strangeness. The voice was part of the physiognomy of her suffering: stuttering, uncertain, with a hint of deadness. This added to her so-called genuineness and was read as further evidence of her integrity. Magical it was indeed that by simply *crying* without prompting at the presence of AIDS babies or victims of land mines she could get the world to say she was another Mother Teresa. Mother Teresa had spend her entire adult life caring, day in and day out, for the poorest and the dying in Calcutta, enduring heat and stench and pain and difficulties of all kinds without complaint, and here is a woman who gets off a plane, shows up before a thousand cameras, reaches into the dirt and utters in a quiet, shaking voice that land mines are a terrible thing, and the world jolts. All she had to do was *be* there and emote: before the camera. Christian to the core, her "morality" was read directly from her passion, from her pain at the world. Thus did she excite ancient desires for religion in a vast public caught in modern life. Max Weber called it the "charismatic personality."

Diana's empathy for those who suffered extreme pain was vital and unrehearsed. Nelson Mandela said it in an introduction to *Diana: The Portrait*,[3] put out by the Princess of Wales Foundation (all proceeds to her charitable causes):

> When she stroked the limbs of someone with leprosy, she did more to break the taboos surrounding that disease than any number of books, articles and health education programmes. When she sat on the bed of a man with HIV/AIDS, held his hand and chatted to him naturally as a fellow human being, she struck a tremendous blow against the stigma and superstition which can cause almost as much suffering as the disease itself.[4]

Her naturalness on the TV and with HIV/AIDS sufferers confirmed that she walked among us while being more than us. "We cannot all be a famous British Princess," Mandela continued. "We can, however, all try to do what we can to insist that every human being is precious and unique." This identification with her was a matter of her own ability to convert personal suffering into gestures of sympathetic identification with others. The public read her that way, as a figure who invited identification of their own suffering with her own.

The pain went way back, to the fate of a girl born one year after her parent's second child, who lived only a few hours. Diana was meant to replace the child loss, but did not and could not. They'd wanted a boy. She always knew herself to be a bitter disappointment in a way that must have happened before her ability to speak and must have haunted her language centers like a torpid ghost. Mothers in mourning for lost children, mothering the new ones in alcoholic distraction, furious, perhaps, abstracted, the burden Vincent Van Gogh felt, and Gertrude Stein, and many others who were born to replace the irreplaceable, repair the irreparable. Diana had a heaviness not true of her younger brother Charles (a boy, and born later). This made her natural joy more sparkling, because more transient and more deeply felt. And it made her a person living in need of grace. Her ability to give it was felt by her public to be related to her need for it. This conversion of suffering into cure was, amazingly, something that came across through tabloid and TV. She was thus marked with grace, fated to carry a saintly aura.

From a British Film Institute book, based on survey research into the Diana phenomenon, here are two reactions to her death:

> I did not have a happy childhood myself and, like her, I have sometimes felt out of control in the past as I've worked my way through various traumas to reach the point I'm at now. . . . I feel a certain respect for Princess Diana that she managed to hang on with gritted teeth almost and deliver a substantial amount of good despite her own difficulties.

And,

> I was devastated [by her death]. She was a very special member of
> the Royal Family and reminded me of so much of myself. I too had a
> loveless marriage—I felt so deeply sad inside but never showed it to
> others. I tried to be happy, I too suffered an eating disorder because
> I was so dreadfully unhappy. Deep down I was crying out for help
> but no one understood. My husband said I was seeking attention and
> had no sympathy.[5]

A concept of Richard Dyer's about the film star is useful here.
Dyer writes:

> The phenomenon of audience/star identification may yet be the cru-
> cial aspect of the placing of the audience in relation to a character.
> The "truth" about a character's personality and the feelings which it
> evokes may be determined by what the reader takes to be the truth
> about the person of the star playing the part.

Treated as victim of her own position, and therefore one among
the multitude of suffering congregants, she was vivified as saint.
She could only have taken on this role because of what Richard
Dyer calls her "star image."[6] This was a combination of feeling and
physiognomy, of royal background and natural beauty. In our time
persona carries moral authority, just as the voice of the newscaster
is increasingly the grounds for believing what he or she says about
the world. Diana's moral persuasiveness was a matter of public belief in her sincerity, and the key to this was her spontaneous physiognomy, her crying, bodily shudder, gesture of drawing a baby
to her breast. I have little doubt that she was sincere. But, if one
wants to ask how the media constructed her, part of the answer is
that through the lens of the camera her physiognomy became the
grounds for her sincerity and was raised to film star power. No one
thought to ask if any of her gestures had been set up (for the camera, for example), if the babies had been planted, and so on. Everyone wanted to believe in the spontaneity.

Her electric physical communication combined with her verbal dumbness made her apt for the silent star image: and hers was at odds with the royal posture where silence bears the authority of exclusion, the power of command. Normally, star image precludes allegory, the reading of personal history as morality play. But in Diana's case the opposite was true. In her anxious twitches, frozen silences, spontaneous gestures of sympathy, personality became parable. The royals hated her for her own body language and for this mantle of morality she wore. Act fatuously and carry a purse, stand dutifully above it all: this is the royal adage for women. Inside the purse there is gin in the flask, which can help. The men are free to consort with their consorts; the women are all stiff upper lip and inner fury. The queen hated Diana for her inability to remain in a woman's usual state of dutiful emptiness or quiet desperation. This was a failure of royalty, of duty, of *Englishness*. Even Princess Margaret had towed the line! Royals should be seldom seen (except skiing in Gstaad) and known exactly never. She could not help but reveal herself; her story reflected in her face like a mirror; her body was knowledge incarnate.

A recent film (*Queen*, 2006, directed by Stephen Frears) reminds us just how cloistered the British royalty had become after the Second World War, how out of date were their frozen demeanors, how disliked for their duplicity, how arrogant were their refusals of public presence. The Queen Mum had been hugely loved for her steadfastness during the London blitz; she drank her way through the bombings yes, but in danger, in Buckingham Palace, in London with her people. Elizabeth was cold, Philip off at polo in the Argentine, neither able to satisfy the public's desire for a cult of royalty. Charles was a schlemiel and a philanderer, a man who would not or could not play by the rules: either give up your consort when you marry or keep the wife in tow. "There were two Dianas," Charles announces in that film, the private Diana and the public, implying that the private was everything the public was not, unbearable, insufferable, a burden. No doubt rationalization, but Charles's remark is also true insofar as the public's Diana was an artifact of the media, known only through it. And she, "The

People's Princess," the one who, first since the war, had given the public what it wanted: a figure worthy of their desire of cultivation, that is, of cult. The film sets Diana apart from the others by showing her only in film clips, TV interviews, photos, contrasting sharply with the dowdy ordinariness of the queen.

And so Diana's public held a stake in her life, wanting its outcome to go one way or another, waiting impatiently for the next episode. News of Diana from the tabloids, her ongoing story, left the public hanging. Every chapter was magnified, turned *episodic* in the manner of television. She embraced AIDS orphans over the TV, and it was read right back into her star image, enhancing the magical magnifying glass through which her next episode was read. A dispossessed star, seeking liberty, seeking solace in a cruel world, active, joyful, depleted, unable to eat: this was the stuff of the greatest show on earth. A soap opera is, after all, an *opera*, although usually not one with high-level diva stars. She was a star without being a diva, tracked in every episode of her life with the expectation of melodrama. This tracking of her life, episode by episode, is the aspiration of television, which holds an audience in suspension over the airwaves so they can't wait for Sunday night or Monday noon to discover if Thorne is going to get married, Tiffany will adopt, or Tony Soprano will get whacked. Reality TV is a new generation of game show taking place on bikini islands covered in flowers and poisonous snakes. Diana's life was the *real* Reality TV, that rare stuff of public obsession over a story that was not a setup (game of getting off the island or entering the secret maze) but really happening. Her story was one large reversal: royal now hounded, princess now outcast, purebred turned talk show confessor, woman scorned, then reborn with Middle Eastern nouveau riche playboy, about to start (what else!) a production company, Euro-princess cavorting with high-class Eurotrash, Euro-princess about to *become* high-class Eurotrash. There is no better daytime drama than reversals of fortune.

She pled her case through the media and before the public and was tracked by the paparazzi, hunted by a breed of journalists as vicious as the hounds the royals themselves kept. "It was an irony," said the Earl of Spencer, that "a girl given the name of the ancient goddess of hunting was in the end the most hunted person of the modern age." Hunted in her Mercedes, speeding away from the stalkarazzi at two hundred kilometers an hour through the tunnels of Paris, her driver drunk, with no police escort—it had been removed some time earlier by the royal family, which clearly wished to dispossess and not simply dethrone her. Journalism prizes the distant, the larger than life, the unassailable, demanding access and accountability, delivering image and judgment. And it prizes especially the star, wanting to revel in her presence, while also collapsing the distance between she and her public. Yellow journalism wants to manipulate, intervene, expose star as object of scandal, prove the politician a porno fiend, show life up as a soap opera, generate (after 911) terror by finding evidence of Al Qaeda in aspirins, freedom fries, and foreign Romance novels. These things, as Michael Moore has taught us, sell. Diana's was the rare case where yellow journalism actually gave her life the sensational ending it wanted: a car wreck increasing news ratings by big percentages.

In turn she came to play the game, reluctantly but well. The silent goddess talked, about her eating disorders, her marriage, her children, her career. Charles, Earl of Spencer, was keenly aware of the flak she'd taken for pursuing this route. In lines from the penultimate version of his eulogy, which he deleted, he addressed the point: "To say she manipulated the media is to miss the point. . . . She had to defend her inner self by bowing to the juggernaut strength of the press occasionally."[7]

Within this melodrama of hunted, huntress, exemplar, she remained, spoke Charles, the same "big sister who mothered me as a baby, fought with me at school and endured those long train journeys between our parents' homes." And yet he describes her in the language of a star, speaking in the plural as one among many. She was "the unique, the complex, the extraordinary and irreplaceable Diana whose beauty, both internal and external, will never be extinguished from our minds."

Such was her extraordinary appeal [he earlier said] that the tens of millions of people [actually 2.5 billion] taking part in this service all over the world via television and radio who never actually met her, feel that they too lost someone close to them in the early hours of Sunday morning [September 1, 1997]. It is a more remarkable tribute to Diana than I can ever hope to offer here today.

In a more academic language Richard Johnson remarks on a similar thing:

Even in her death Diana bequeathed to others the opportunity to grieve for ungrieved bereavements of their own. Misrecognized by critics as manufactured or sentimental, this mourning [by the public, on the occasion of her death] was . . . [genuine] public grief and personal loss, magnified as much by the intimacy and extent of Diana's social connections (face-to-face and mediated) as by modern . . . media.[8]

The word *cult* is not out of order for a woman who could generate this intimacy, gestate her "particular brand of magic." The only thing absent from the story is an Andy Warhol to paint it in duplicate or triplicate, to canonize her in acrylic, make of her a Marilyn or a Jackie. When Tom Fleming spoke of Diana's first overseas assignment, "representing the queen at the funeral of a beautiful and much loved princess who had died some days before in a tragic car accident," and went on to say: "Little did I imagine that fifteen years later to the month I would be watching the arrival of a simple coffin draped with *another* royal standard, bearing the body of that beautiful and much loved young princess of our own country, killed six days ago, in a tragic car accident," he was pointing to symmetry between these stories. Each seemed to mirror the other, as if, through some equally particular brand of magic, each had taken on the features of the other, virtually becoming it. Diana was no film star, yet she took on the mantle of that stardom before the camera. She was a woman who had never acted in a film, but seemed to carry the aura of the star through some special baptism. Grace was born of no royal blood, assuming the royal posture with fairy-tale perfection when she married Prince Rainier, and this because she was already queen of the cinema before going on to become princess of the five-star hotel and the gambling casino. She was goddess before becoming princess. Each had a face of the classical ice goddess, a face also more than human in its way, revealing deep passion, contentment, cunning in Grace's case and moving desperation in Diana's. Each lived in a world of seclusion and fast cars, away from the media and yet wholly of it. In each the distant mantle of royalty blended with the distant mantle of the film star. These are also the stories of Marilyn and Jackie, those of a particular genre. Half fairy tale/half woman-on-the-verge melodrama, these beings exist between real life and the netherworld of the camera and in death become radiant icons in the museum of the public's imagination. Diana takes on the attributes of Grace (movie stardom), and Grace Diana (royalty, pedigree). Diana's Grace and Grace's Diana.

This is our world, the world of Marilyn on-screen forever, Sugar dazzling in a tight skirt, spontaneous before the camera, good for the shareholders, funny, vulnerable, sunk in passion for saxophone players, bursting with soft, Veronese fleshiness, her torso a jazz riff, her eyes pools of light. And the world of Marilyn off, screaming in quiet desperation, swallowing the pills and not making it to the telephone. Billy Wilder had to do over a hundred takes of the scene on the yacht where she tries to "cure" Tony Curtis of his "ailment," and she barely made it without collapse: neurotic, desperate, consumed with insecurity, always about to freak. She left the bulk of her estate to her therapist (to pay for the sessions she's missed on account of being dead?), nothing but memories for her glamorous, loving, impossible husbands. A persona, a set of films, an iconic existence continuing beyond death, made alive through it: this gift for her public, a gift largely not of her own making.

Then there is Jackie with her Camelot existence, the assassination, the indelible image (John-John in his sailor suit, bidding his father goodbye), the presidential funeral—so drenched in media that she seemed permanently suspended between life and art. Surrounded by glitterati and Eurotrash, needing to get away from the ironclad grip of the Kennedy clan, Jackie resolved to raise her children abroad, to find another life, which she did on a yacht in designer swimsuits and elliptical sunglasses to the detriment of a real diva (Callas) and another melodrama with another death. She lived the second half of her nine lives at couture runways, hidden from journalists, in pursuit and in countersuit, filing lawsuits to retain privacy, becoming a businesswoman in New York. But, wherever she went, Jackie had orchestrated the presidential funeral and remained permanently within it, as if her life went on but also carried the aura of that moment in the form of a media halo. It haunted her, that halo in which she would always remain, laying the wreath on the coffin, forever in the public's imagination, funeral and photo endlessly shown again, an icon that would never go away. It is hard to live in the present if you are also an icon suspended in time before the gaze of millions (within their gaze). Grace did not seem to mind the way the public projected her past onto her present, her film star status onto her queenly virtues. She seemed in love with the camera, assured in her narcissism. When she relinquished stardom for another life and transported its glamour onto a European stage, she was happy enough to be seen as a star in her new role in the casino royale of Monaco. Perhaps she didn't care anymore.

These women lived lives as people and also as personages or icons. The personage was a thing largely uncreated by them. They existed here and elsewhere (there, in our imaginations, on screen, in the past, in the media bank, the aura of lost things). They found ways to accommodate this or were killed by it, depending. Whatever happened, it simply fed the public appetite for more of the persona.

The pairing of Grace and Diana is hardly one to have escaped notice, but, ironically, little has been written about it. The books I've read tend to be straightforward narratives, with a certain frisson of gossip thrown in for the "female reader."[9] The persona does not, of course, require any books. The night before I taped the Diana funeral from what was then my house in Durban, South Africa (her funeral was the third most watched media event in the history of South African television), I'd been busy screening Grace Kelly in *Rear Window* (1954, directed by Alfred Hitchcock), thinking about Liza's remark about being studied "like a bug under glass." Then the next morning the Diana funeral, the result of a butterfly trying to escape the glass.

Bugs and butterflies: the old studios (from the 1920s to the 1950s), understanding the public's appetite for cannibalization, took it upon themselves to build up the star through a combination of withholding her and offering her to the public. The studios understood that too much publicity, too much viewing of insect under glass, will demean stardom by turning it into the daily fare of celebrity. The star in the old sense was about unapproachability. The theory was he or she should be present to the world primarily on-screen, not off. Such an exalted being is perforce also a celebrity, the result of film stardom and careful studio buildup. When he or she made an appearance at openings, charity balls, or was caught on film shopping the public gasped. Craving anecdote, scandal, images of ordinariness, the public, given free range, consumed the star, got closer and closer to her. What the studios grasped was that, unchecked, too much closeness would be a failure through victory, since the aura of her distance, which prompted the desire for intimacy, would be lessened, reducing her star value and producing final disappointment. Scandal can wreck a career, too much familiarity can make it prosaic. Given this, the familiarly craved by the public was provided an iota of satisfaction, which did fuel desire for more and keep the star in the public eye. This public, the studio understood, also did *not* want its stars to be people of ordinary life but, instead, filigrees of light. In this frame of mind the public desired proof of the star's flesh-and-blood ordinariness, only to establish the mystery of transitions: the fact that stars are real people who appear in transformed form on-screen. Thus stardom existed in an uneasy relationship with the celebrity it also was, for the terms of stardom were in opposition to those of celebrity. Marilyn's celebrity came in dribs and drabs: through what we knew of her marriages, through tabloid reports of her consistent failures (drugs, alcohol), through her notorious difficulty in the studio (finally she was fired from a film). But she became iconic mostly through her death. Had she gotten to the phone, her story would not have taken the turn it did, the melodrama would not have come to conclusion, the icon would not have properly formed before the public. These details of her life were enough to allow her melodrama considerable public attention, but she was also shielded, kept apart from her public, thus retaining star quality.

"The film star aura was . . . built on a dialectic of knowledge and mystery," P. David Marshall writes. "The incomplete nature of the audience's knowledge of any screen actor became the foundation on which film celebrity was constructed into an economic force."[10] My point is that this incomplete knowledge was also internal to the cinematic power of the star, internal to what *made* that person a star in film and *maintained* that aesthetic role. And so the older studio star was partly shielded from celebrity.

However, since the 1950s, with the ascendancy of television, stardom and celebrity have come to form a new alchemy. Even if the studio still existed as a machine of production, it would be unable to control television access to the day-to-day live coverage of the star. The star is someone given to us only through a particular kind of story (her films). She is distant (transcendent), her screen presence is inspirational and yet particular, detached from our lives. She exists mostly, perhaps exclusively for her public, through fiction. The celebrity is someone whose *life* is our interest. Television began in the 1960s to mass produce the star: "Star of the day, who shall it be?" began the theme song of Ted Mack's *Original Amateur Hour,* as if every child with store-bought ruby slippers and a silver clip in the back of her hair could end up a Shirley Temple. This mass production divorced stardom from talent, not to mention that *je ne sais quoi* that is the peculiar, unrepeatable screen presence of each. Stardom then increasingly became an advertising concept, something attached to what is, in effect, celebrity.

Ironically, a media consumed with the raising of all things (all celebrity) to stardom is one that finally turns the star into a simulacrum. This was understood by Robert Musil in *The Man Without Qualities* with respect to the concept of genius. Ulrich, the novel's main character, comes to realize that he can no longer think of himself as a "man of promise" when he reads an anecdote about a horse. Musil tells it like this:

The time had already begun when it became a habit to speak of geniuses of the football-field or the boxing ring, although to every ten or even more explorers, tenors and writers of genius that cropped up in the columns of the newspapers there was not, as yet, more than at the most one genius of a centre-half or one great tactician of the tennis court. . . . But just then it happened that Ulrich read somewhere . . . the phrase "the race-horse of genius." It occurred in a report of a spectacular success in a race. . . . Ulrich, however, suddenly grasped the inevitable connection between his whole career and this genius among race-horses. For to the cavalry, of course, the horse has always been a sacred animal, and during his youthful days in the barracks Ulrich had hardly ever heard anything talked about except horses and women. That was what he had fled from in order to become a man of importance. And now . . . he was hailed on high by the horse, which had got there first.[11]

Such exaggerations of language are (perhaps) even more notoriously rampant in our own times, when Stephen Soderbergh's film *Kafka* (1991) can be called "a mega-masterpiece" by one critic, as if being a mere masterpiece (like Kafka's work itself) is no longer good enough for the terms of mass consumption. Genius increasingly becomes reduced to celebrity, and both become marketing adventures in the theater of presentations. So one ends up with the landscaper hired by my West Hollywood condominium association in the 1990s because he was "Tom Cruise's gardener" (carrying the whiff of Tom Cruise's celebrity by osmosis). This whiff was the criteria of excellence, the job reference. It is the stuff of Woody Allen's *Celebrity* (1998), a mordant film about a New York now become Los Angeles where everyone bows down to the person who reads the weather on the local news channel because he too is part of television land.

In this world there is little place left for the aesthetics of the aura, of the film star.

David Gritten, a film writer and critic, says in a fine book that

the trappings of celebrity can look like a career structure in themselves. Young people, asked what they want to be as adults, sometimes reply, "Famous," with not a single thought to what kind of endeavour it might be attached. And in fairness, why should they think differently? Ours is an age in which fame no longer necessarily depends on achievement; our TV screens, newspapers and magazines are filled as never before with people who seem famous just for being famous. (The American historian Daniel Boorstein first coined this idea: "A celebrity is a person who is well known for his well-knownness.")[12]

Celebrities are well known for their own well-knownness: a provocative turn of phrase. The celebrity system runs on itself; the celebrity is valued in virtue of mere participation in the system. Put a slightly different way, celebrity is increasingly a media effect combined with public appetite. These together set its "market value." Content becomes determined through the simple fact of participation in the relevant medium (television land). This wows the public. The medium is, in the famous phrase, the message. And it remains a message because the public is ready to accept it, like voting for a mannequin instead of a president. The media that provide the message are three: television, tabloid, talk radio. They are the pure form of commodity value, the system that sets the value celebrity. Celebrity is like money: a "currency" whose value waxes and wanes through these media forms. This is, of course, too simple, since many different factors generate celebrity, including talent. But it is more correct than one would like to admit. "We're scouring every facet of American life for stars," said *People* magazine editor Richard Stolley in 1977.[13] "We haven't changed the concept of the magazine. We're just expanding the concept of 'star.'"[14] Most are talentless, without character, incapable of taking on any of the old—(dare I say it?) more profound and enthralling—aspects of stardom, lives with one-liner stories. Even in film the star is an increasingly absent phenomenon, while the celebrity actor multiplies. Her marriages, children, religious affiliation, sexual tastes, drug addictions, childhood abuse, infringements of the law appear daily on television and in the newsprint. Angelina Jolie and Brad Pitt auctioning off the rights to the photo of their adopted baby in order to donate the proceeds to charity: these things point to lives lived for the television—and *of* it.

Celebrity is an expanding market, a robust economy. *People* magazine makes a person into a persona, whoever they are. In this system a person becomes a celebrity by being in constant circulation, and this circulation (being endlessly talked about, endlessly available on television) sets the celebrity value. It is destructive to the *distance* of the star, because her stardom depends on a modicum of public unavailability. She is simply out there too much and too often to retain the halo of the silver galaxy. Celebrity and stardom are, in contemporary life, largely at cross-purposes. This the old studios well understood.

What is unique from the point of view of aesthetics is the rare phenomenon in which stardom and celebrity alchemize, rather than their being in an antagonistic situation with celebrity often *burying* stardom. Diana's stardom is not buried by her celebrity but complements it. She is, I think, new to history and extremely rare to it. Before her were Jackie and Marilyn and Grace. They are creatures of film, of television, and of media, whatever else they are. Their day-to-day goings-on are of the soap opera variety, while they continue to bear the hallmarks of transcendent, unapproachable stardom. Their aura is a unique one, reducing neither to film, nor television, nor any other single aspect. The celebrity market is a threat to the star icon, but her being out there in the media is also constitutive of her identity as a star icon because only through television and tabloid could her melodramatic life be watched and public absorption in it be created and maintained. How she retains the mantle of stardom while simultaneously living in soap opera TV and tabloid is one of the great mysteries. That she does is the subject of this book.

These star icons are fundamentally new in the history of aesthetics. Marie Antoinette had a cult flutter around her; Sarah Bernhardt was a star *ne plus ultra*. Queen Elizabeth, when young, was herself a media figure, her coronation watched by millions. The famous have their history, including a history of evolving means for becoming famous. What is new about Diana and company is the media and its cult of celebrity through which they rise.[15] By the media is meant more than the tabloid papers (which have been around since the nineteenth century), but also film and television. Without a proper understanding of these—and, in the cases of Diana and Grace, royalty—the aesthetics of the icon cannot be grasped. These are aesthetics that transpose themselves onto the icon, graft onto her image. Diana never appeared in any film and yet carried the halo of film star—something, David Lubin has demonstrated, that already pertained to both Jack and Jackie during their time in official Camelot, and even at the moment of his assassination.[16] Diana's stardom is a portmanteau placed on her in the form of an aesthetic transfer in the public imagination from films to her. In her persona the distance of royalty, the classical beauty, the natural physiognomy allow her to become, in effect, Grace Kelly the film star, without ever having been on screen. Lady Di is perceived to carry the attributes of Grace Kelly, her screen presence, while Grace Kelly glides effortlessly from film stardom to princess of the realm, best in show, becoming the perfect royal, even if her kingdom was an ersatz one run on roulette. Diana's stardom survives the television melodrama of her life, shines through it, lending it a special, aesthetic grace seldom if ever found in an actual television series. Each of these aspects (stardom, TV image, melodrama from life, royalty) harmonizes with the other, painting the whole in a particular, unreal cast. This is the mystery. Only a few have the right physiognomy, the right story, at the right time, with the right media attention, to simultaneously live in the glow of the film star (whether or not they have acted in movies) and the soap opera of television and tabloid. They seem to be created by that great casting agent in the sky, whose ways are ever opaque to the system, whose hand is invisible to the market.

I think this particular icon quality is, in our current world, vested mainly upon women: a feminized role demanding the quality of the star and the melodrama of a life lived passively, receptively, operatically, which is a gendered life. It is a role closely linked to concepts of hysteria and visual objectification. JFK was in his own way a kind of icon, for there are many *kinds* of icon, all with star quality matched by celebrity and an exemplary status of some kind. For some, John Wayne was an icon (exemplary of strong stride, tough masculinity, a slow drawl, and a decent conscience). He could do vulnerable, at least in love, at least with the camera of Howard Hawks leading him on. His icon status was as an indomitable settler, settler of land and argument. For others, Paul Robeson, big and wide as Wayne, football player turned lawyer, Old Man River rolling along as Communist, emperor (Jones) if not king, star in all things, never less than fully human, never free of the painful drama of black history, always battling the hand that wanted to keep him down, a prince among men. The younger generation has its own icons (Johnny Depp?). Of these varieties, the kind I describe is particular and seems to last in the public imagination beyond the others, for it is a kind that catapults the star into an even more distant galaxy of perfection by juxtaposing her stardom with the melodrama of her life. Some men approach the type: JFK had screen presence and star quality, cult following, and a load of trauma dogging him (the death of his older brother in the war, his intractable pain). And his assassination solidified his place in this particular firmament, although I think, it solidified Jackie's place more. Were it not for the assassination, JFK's icon status would have been different: in spite of the war and the load it made him carry on his back, the man was too happy, too conniving, too much in control of things, he was too much the Don Juan. His assassination brought the melodrama home. Elvis is perhaps the greatest male example of the kind of icon I've in mind: with an electrified physiognomy that hit from the belt and reverberated in the voice, a love of celebrity that was out of control and a gradual decomposition into drugs, isolation, a retinue of personal servants, physicians, and bad family relations that are the stuff of a Douglas Sirk film. He qualifies, I think, but if he does qualify it is because the melodrama had feminized him, because the bright colors and the acrylic hip hugging suits that bulged his masculinity also made him a woman, like he was overdressed for life and ready at the first glance to weep. This appliqué of femininity to the ultra male is a lip-syncing lipstick that sends gender haywire. Evidently a minimal condition for the icon is that he disrupt the canons of masculinity through his role as a figure of melodrama. To bring the point home: Lassie could be an icon, but not Rin Tin Tin.

A great deal has been written about the objectifying gaze of the cinema lens, which craves the study of stars like bugs under glass. It is systematically a projective gaze, in which femininity is vested onto the woman through the encaging lens of the camera.[17] It probably remains true that women are as a whole less free, less idiomatic, less independent than men, which is why they've greater liability to the winds of fate and also a public readier to read their lives that way. The role demands this picture of liability, of being perpetually out of control, a woman on the verge. Moreover, this is a persona whom the public want to eat alive, at the same time preserving like the bug under glass, and such control over a being, even over her image, is, again, vested more completely when the image is that of a woman. The gaze "womanizes." And so most icons of the melodramatic kind I've in mind here are women. The other side of this Madame Butterfly diva role is that we are profoundly sympathetic to the person in the female *persona* struggling to be a person, to the human drama of it all. Having imprisoned the dame in her gilded cage, we then seek to set her free. We find it moving to watch her live or die doing it.[18]

There Is Only One Star Icon (Except in a Warhol Picture)

AS ABSORBING as the Diana story is, this is not a "Diana" book. I was no part of the "Diana cult," nor do I have any stake in praising or blaming her. I am not a British "subject," nor am I a card-carrying member of any star society. To me the only true royals are the Marx Brothers—especially in *A Night at the Opera,* since I am a fan of opera, both comica and seria. That this silly, overbred, haunted girl became an icon before the masses is as unlikely as the Wagner opera *Parsifal,* which is so seria that it ends up comica, being the story of a pituitary case from the Rhineland who stumbles through his lobotomized life to become a hero and save

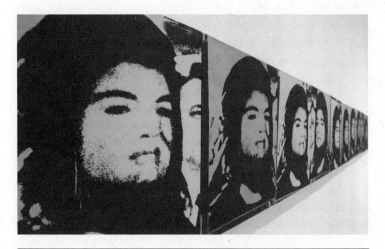

FIGURE 2.1 Andy Warhol, *Jackie Frieze.* Photograph by Justin Lane, Corbis

a middle-aged Germanic cult from its blood and gore. This is interesting, since *Parsifal* is myth (although it did end up history around 1933), while Diana is the real thing (super-real given the role of the media in making Diana a figment of light). I am no fan of cults, finding them, like so many intellectuals on the left, deluded and dangerous. But I should also say I write as someone with a lifelong fascination. I find myself stunned by the star—a fixation I cannot get past. This comes from way back, a childhood watching movies home on a Sunday, eating TV dinners with my parents and brothers in a family that would, soon enough, generate its own brand of melodrama. My mother was always silent, deferential, glad to be of use; my father wept, wept before the old Hollywood movies replayed on television of Louis Pasteur (Paul Muni) fighting the French medical establishment, Victor Laszlo (Paul Henreid) beckoning Ilsa Lund (Ingrid Bergman) to a life of comradeship without love because, even though she didn't understand it yet (so said Richard Blaine, aka Humphrey Bogart), the problems of three people don't amount to a hill of beans in this crazy world. He wept because Charlotte Vale (Betty Davis) found romance on a cruise away from her crushing mother in the arms of the same Henreid (who evidently got around in high-class circles, but then he was a Viennese aristocrat). And he wept when the Morgan family emerged from the Welsh mines grimy in singsong because soon the valley would no longer be green, in spite of the good Walter Pidgeon. His was a boy's world of heroes vanquished, mothers ever faithful, the new land ripe with invitation. At night he had dreams in which he was Atlas, holding up the world; his own mother's brutally uncompromising voice weighing him down like an angry god. We all knew that the intensity of his emotion rose above its subject, as if he himself lived in the cinema screen, believing himself Gary Cooper walking down the streets of a deserted town at high noon, alone and afraid, his upper lip twitching, his eyes anxious and narrow, his courage that of a World War II veteran invading Omaha Beach. This knowledge made us uneasy, so that we were always looking to restore him to gaiety and ourselves to cheerfulness through some kind of play acting. We made movies, we three brothers, out of the stuff of Hollywood romance and acted them out for his benefit. His response would be to overwhelm us with tearful gratitude, which caused us the anxiety of children upon whom elders too much depend. My father was a man with a river for a soul and the boxing gloves of a John Garfield. You never quite knew if he was reciting Shakespeare or punching you out. He could be diabolical, sadistic, evil, but at these times we were touched, not infuriated. For me black and white still carries his tears; Kentucky Fried Chicken the taste of childhood. His contorted face was funny, and melancholy, and that Janus face is to me the face of the star icon. In the love of the stars my brothers and I bore was always hidden the melodrama that would later explode, causing each and every one of us to wrench in agony, like Diana before her "mam." Oh, and I should add one thing: my father was a fashion designer. In our family what you wore was more important than how you felt. My mother dressed like a Barbie doll. She was her own kind of icon: cheerful, warm, plastic, sepulchral, powerless.

I am trying in this book to understand the icons I loved before I was old enough to grasp the force of this attraction, this pantomime drama. These icons captivated me in the way Shirley Temple captivated a different kind of person a generation before me, an immigrant boy from Pittsburgh with a talent the size of the silver screen, an artist who understood in his bones that with the icon melodrama is everywhere—and everywhere concealed (which is why it is melodrama). I mean Andy Warhol. Why not say it? I am totally ambivalent about this phenomenon, which is I think the way a lot of people feel about it. I write as a lover of this genre, but also someone incensed, repulsed, unable to sort out how this bizarre form took hold of an entire culture, convulsing it, convulsing me before I was old enough to be able to describe it. I believe I am in these respects an artifact of my times.

A number of books have already been written about film, television, tabloid, about star, celebrity, persona. There is a Diana industry out there ready to elevate, bury, and exhume the body and then bow down before it, extract every last drop of gossip, malfeasance, malady, or monstrousness in the name of pronouncing it "with us forever." The latest book, by Tina Brown, is an acerbic and lucid story of life, love, and celebrity exhaustion, remaining within the circle of lunches-I-had-with-the-princess, who was, as always, the main course.[1] The book has all the virtues and limitations of Brown's old magazine *Vanity Fair*. As for the latest book about Grace Kelly, it is peppered with such scintillating and adept turns of phrase as "Accustomed to having his every whim fulfilled and buttressed by his million-dollar-a-month trust fund, Ricky di Portanova had always lived a gilded existence."[2] I hit rock bottom with a book comparing Princess Diana and Grace Kelly by someone who turns out to be a physical education teacher at a Florida high school. There is nothing wrong in that; one good physique deserves another. But on the back of the book the writer described himself as a devout Catholic with children and grandchildren, which I took to be his way of saying "Listen, Jack, I may be drawn to these babes, but a drag queen I ain't. If you want to get physical, then let's draw our guns now, bub" or words to that effect. There are scholarly books ready to claim that Diana didn't exist at all, but was purely the artifact of the British tabloids (perhaps she really was a fortune-teller from the sixteenth century that the media dug up), others telling us that, while she was real, her audience's feelings for her could not have been because they were mere constructs of the media (whatever that could possibly mean). Two things are clear: her audience loved her and they loved an artifact floating above, in the media, in their own imagination. How one sorts out the aesthetics of that is anybody's guess.

Books on Diana have yet to focus on the aesthetic features of the media that composed her like a hair stylist, on the larger landscape of icon, aura, cult into which she appeared, on the doubleness of her being as person and floating opera directed by some on-the-job media conglomerate. Diana was the creation of a multiple aesthetic, the object of a divided public desire. She was marriage material for Charles because of her lineage; she became a public icon by virtue of her physiognomy, her story, and the culture of tabloid, television, film. She was a creature of film, television, tabloid, and consumer society as much as of royalty. Without the star system in its British incarnation, there could have been no Diana out there in the public eye. These elements conspired to generate public desire around her. Her public wanted her to live in the aura of transcendent unity (the royal/film star), but also wanted her life to remain rocky, unresolved, melodramatic so they could hang on for the next installment (the TV/tabloid goddess). The public saw her as inextinguishably radiant, while also in a state of semipermanent decomposition. They projected unreal perfection onto her image, then read her life story accordingly: sympathizing with her eating disorders and her self-inflicted pain, reveling in the whopping good story of her husband's infidelities, remaining glued to the details of her divorce like viewers of *The Truman Show* (1998, directed by Peter Weir, screenplay by Andrew Niccol). They identified with her pain and with the grace they themselves vested her.

At the core of the Diana cult was this public ambivalence about grace. Her aura was seen to radiate it, but she was also loved because this aura eluded her. Cary Grant (who suffered depression) once remarked: Everybody wants to be Cary Grant; even I want to be Cary Grant. That he was from the English lower classes, one Archibald Leach, who arrived in Hollywood an acrobat from the stage to become Cary Grant in name, and in persona, while constantly eluding the aura of that personage in his own life, is a mirror of Diana. Except that it is not even clear she wanted to be the auratic being the public created, indeed it is not even clear she understood its terms.

Diana was a creature of trauma vested with transcendence, and the public's ongoing desire for her to remain both is what distinguished her as an icon. As I've said before, icons differ from mere celebrities by virtue of their star quality—and from other stars by virtue of the way the public reads their star quality against the narrative of their life. I've said there are a number of varieties of icon, from the John Wayne type to the Mick Jagger can't-get-no-satisfaction variety. All are tinged with star quality, celebrity driven and exemplary to a generation (although evidently not in the same way). These features make them icons of whatever type. The distinguishing feature of the Diana type is the doubleness of the persona:

transcendent yet traumatized. She, the star icon, is an instance of the public's desire for a world made whole through beauty. Theodor Adorno would say that modern times have dehumanized life to the point where the bounds of sense and evaluation have been burst: we can no longer find a way to speak, understand, evaluate what we have done to ourselves and become. There can be no poetry after Auschwitz, he famously remarked. Against this claim that authenticity consists in the refusal to speak, given the failure of language and the preposterousness of redemption (or even acknowledgment), the cult fixates on those whose raw terror can be made lyrical, rendered beautiful. This is a way of lying to the world about what is at its core, a similar lie to that which gives rise to religion. The aesthetic of the cult proclaims that redemption is possible even in the dregs of ongoing despair. Rather than acknowledge that suffering is waste, emptiness, lack of meaning, the cult turns to the suffering of the star icon, makes her aura into something transcendent, identifies with that transcendence, and thus practices a view of the world in which reconciliation with suffering becomes imaginable through her, in which the initiates' own suffering becomes mysteriously elevated. Around Diana's suffering, the public misrecognizes itself, as Jacques Lacan would put it, falling into the fantasy that its own lives are reflected in her halo and so its suffering is also, like hers, referred to beauty, made beautiful, redeemed. She lifts us up because she is brought down while remaining in our outer galaxy. Through misrecognizing ourselves in her we become transported above ourselves. This is the gaze of art that sublimes. Her aura allows for the misrecognition of wholeness, of reconciliation between suffering and beauty. Diana becomes for the public a sign of its own longing for grace, of its own image of grace already achieved and hovering above heads like a halo. The Diana identification is like a reconciliation with Christ screaming to death on the cross, which, rather than being abhorred as mere waste, cruelty or the sadistic politics of empire, is taken to be the sign that redemption for all is possible.

Trauma and transcendence. That picture of Diana, sitting in front of the Taj Mahal on the India trip, dejected, her head cast down, was a first sign of something seriously wrong. Diana was later obsessed with Mother Teresa, following her all around Europe, wanting to sit in her shadow, to become her. Diana, believing herself bereft of life, impoverished in every way, the poorest of the poor. When Diana appeared before her public haggard and emaciated, this excited public sympathy, but also the sympathetic nerve, allowing her public to refer its own pain to her beauty, misrecognizing themselves as similarly beautiful in suffering. She became their self-fantasy, the beautiful soul who catapults their own miseries into another, better realm.

This is dangerous enough, if also moving, since it is an attitude that ends up removing consciousness from recognition of the true nature of its suffering, not to mention the social injustice that partly gives rise to it. But there is an even deeper delusion built into the aura of purity and grace projected onto the fragile Diana. A refusal of the public to recognize its own sadism, its own desire for blood, melodrama, violence, to see that its wish is (at one level) that the Diana story never be resolved so that it can continue to enjoy this "weepie" on the telly. Secretly desiring reconciliation, they also desire more opera, they are a Roman public fixated with lust for death as the gladiators are about to fight—they can't stop. The media generates this through its obsessive presentation of her, but public desire for execution, gore, blood certainly predated the making of television out of it, as any story of the history of public execution will tell. The public refuses to admit that the more Diana remains in search of a grace that eludes her, the more they like it.

And so around the cult icon (Diana) an ambivalent attitude toward her grace is played out.

Now most every form of celebrity is an object of some sort of public ambivalence: When a celeb goes off the rails, careening into dope, alcoholism, or a pedestrian, the public feels personally betrayed and genuinely outraged. How dare you do this to us, you who have everything on account of us? We shall pulverize you! As Leo Braudy puts it in his book on fame, "Modern fame is always compounded of the audience's aspirations and its despair, its need to admire and to find a scapegoat for that need."[3] But this ambivalence is played out in a unique way for the star icon, whose very appeal consists in her combination of glow and pain. (Braudy's book does not distinguish between types of contemporary celebrity, but is, rather, about the long history of the subject of fame, beginning in ancient Greece). This double interest in the glow and the pain is the source of the icon aesthetic, the Diana, Jackie, Marilyn, and, the Grace Kelly aesthetic. The star icon in general reveals the human in its range of characteristics: good, bad, ugly, desperate, fixated, blood lusty.

It is a glow and pain focused around the body of the star icon. Braudy puts it elegantly in a description he gives of Marilyn Monroe, whose life he calls

[a] virtual allegory of the performer's alienation from the face and body that are nominally the instrument of her fame. Like the young Elizabeth Taylor, Monroe was both child and woman, to be nurtured and to be desired simultaneously. The sexual lushness she projected went hand-in-hand with the human impression of vulnerability. Wearing a body that was the object of the fantasies of countless others, she felt herself to be empty and so married two sensitive men: first Joe DiMaggio, the publicly certified athlete and gentleman; and second Arthur Miller, the publicly certified wise man and writer. But neither could fill the sense of incompleteness she had, which was as responsible for her public appeal as for her personal failure.[4]

Braudy ends this description with language close to my own: "If stars are saints, Garland and Monroe are clearly among the martyrs."[5] The genre is cult forming in the case of a Judy Garland and downright ecstatic in the case of a Diana.

It is a genre defined by these paradigm cases: Jackie, Marilyn, Grace, Diana, Judy et al. Some are actual film stars; others have this quality grafted onto them (by dint of Camelot and the like). There are intermediate cases between this genre and adjacent ones (JFK is an example, Elvis, I think, another, both of whom fall halfway into the star icon category and halfway into a more male/erotic type). My opening analytic move in this book depends on the intuition that these paradigm examples belong together, requiring a special kind of story. (If the reader does not share this sense he or she will probably find my book most unsatisfying, since this is for me a starting point I cannot otherwise prove than by trying to paint a picture of these dames that emphasizes what is distinctive about them considered as a gang.)

They live in Andy Warhol's universe of the cult icon: a beautiful, cruel universe. Warhol's work joyfully embalms the star while disposing her in flat indifference without aura. His work takes pleasure in her suffering (it is sadistic), but also deeply invested in her (it is empathetic). Above all, is it underexpressive, giving in to neither of these stances quite, instead projecting a deadened, bored, consumer stance toward the star, who is, after all, a mere item in the panoply of product images anyone can consume and, when finished, move on. Warhol's early work from the 1960s dedicates itself to a small and elite collection: those few he recognizes as having icon value. Later he will become more interested, I think, in a department store of celebrity types from Chairman Mao to any rich New York collector who will pay him the 20K to have his portrait painted. In the early 1960s he confined himself to Jackie and Marilyn of the star icon variety, Elvis and Troy Donahue of the more masculine/erotic type. These are obsessively repeated.

Braudy points out that repetition is a source of the glow of fame: "Famous people glow . . . and it's a glow that comes from the number of times we have seen the images of their faces, now superimposed on the living flesh before us—not a radiation of divinity but the feverish effect of repeated impacts of a face upon our eyes."[6] This is exactly right: repetition makes the image of fame glow. But the opposite is also true: image repetition dulls the glow, leading to current "Diana exhaustion," past "Marilyn weariness," the desire to change channels and change icons. It is this doubleness that makes the fate of the icon robust but also precarious.

Repetition also hides melodrama in the pseudo-mechanical forms it generates, as if the star icon's image were hiding from the world in Warhol's dark glasses—the dark glasses of a man who viewed the world from the safety of a moviegoer. Ours is a society where depth is continually converted into surface, emotion into consumer choice. Pain becomes Oprah-speak, illness choice of medication; mortification is staved off in the aerobic palace, destruction concealed through the excess of images that reveal it every day on the TV. Susan Sontag famously said that one photo acknowledges suffering while an endless parade of them deadens response to it. When Benetton can feature African AIDS victims in their ads as a way of selling jeans and T-shirts through "moral sympathy," then the world runs on aura. Ad agencies could not have imagined running these campaigns were the public not ready to invest nearly anything with a halo and then go out and shop for it as if it were just another in an endless line of products. That these ads were pulled proves the public actually does have a moral limit somewhere, but it is pretty far out.

Diana is there to break through this Warhol consumerist indifference, but also to be consumed in exactly the same way. The life of the icon is an adulated life, but it is also life as a mere consumable. Like any consumable, her icon halo might at any point be reduced to mere image product; the public might decide it had consumed enough of her and trash her image or, more likely, let it recede into indifference after her fifteen minutes of stardom. Her sustainability as an icon is always at issue, under threat. In this she is different from the art in a museum or the violin concerto in a concert hall whose career has stood the test of time in bourgeois culture and is not, as yet, reducible to a consumer item, although the iPod may be doing just this. Diana was prevented the fate of immediate trashing because of her English status as royal, Marliyn because she remains to this day effervescent on film and because her death still satisfies public appetite for melodrama and has now become "iconic." Grace Kelly only acted in five (near perfect) films; then became fairy-tale princess before the car spun out of control. Were Diana still alive, no longer royal (except in lineage), long away from Buckingham, married to Dodi Fayed and producing schlock movies in the production company they wanted to form, movies featuring the Spice Girls and Mr. Bend it like Beckham himself, were she a contestant on *American Idol* or *Big Brother*, a fixture at the ready-to-wear and your occasional falling-down drunk, she might have slipped from halo to another celebrity product, another has-been, another piece of designer Eurotrash. We cannot know what would have happened; we can only say that even the halo of the cult icon is fragile, given that it might at any point reduce in T-shirt value. This is part of what Warhol wanted to demonstrate about art—that a Brillo Box and a work of art are fungible, each so close in kind to the other that in our society the one might at any point reduce to the other because it already wears the other's shoes—pertains to the aesthetic value of the cult icon. Warhol, in thinking he demonstrated this, proved the opposite, since his works of art are undoubtedly works of art rather than mere advertisements, or advertisements at all. Yes they are commodities whose value in commodity culture is that of image value. But their rarity is the rarity of Diana and Marilyn: that of a thing of beauty speaking melodrama and in many ways empty of content that is not yet reduced to Benetton.

It is worth noting that Warhol's own epiphany about consumer society took place (according to him anyway) during a car trip he made in 1963 across country, from New York to his exhibition at the Pasadena Art Gallery (now the site of the Norton Simon Museum):

> The farther west we drove, the more Pop everything looked on the highways. Suddenly we all felt like insiders because even though Pop was everywhere—that was the thing abut it, most people took it for granted, whereas we were dazzled by it—to us, it was the new Art. Once you "got" Pop, you could never see a sign the same way again. And once you thought Pop, you could never see America the same way again. The moment you label something, you take a step—I mean, you can never go back again to seeing it unlabeled. We were seeing the future and we knew it for sure. We saw people walking around it without knowing it, because they were still thinking in the past, in the references of the past. But all you had to do was know you were in the future, and that's what put you there. They mystery was gone, but the amazement was just starting.[7]

What Warhol saw was a landscape of signage, a new American landscape where ad and celebrity ruled, rather than water, tree, and mountaintop. He loved the mind-numbing homogeneity of this culture where one thing increasingly becomes the same as all others by virtue of its sameness of circulation (as image). But his actual art focused (at least at the beginning) not on the mere celebrity, indistinct from all others who are well known for their well-knownness, but on the star whose star quality allowed her to become iconic. He was the Russian painter of religious icons for a world of Bloomingdales. His velvet was not that of the religious garment but the velvet underground, which he transubstantiated by embalming it as art. This stuff of images, the new form of American (and global) production, became the source, and the tomb, of the star, turning her (in the right melodramatic circumstances) into an icon. Her value was tripled: her own life, her place on screen or in the endlessly repeated historical moment, her recreation as image-icon by society and painter. This triple life could only have been made possible because of the movies, television, the tabloids, and because the art world shares Diana's fate in crucial respects. Warhol understood this too: a work of art is not far from a star icon; the public desire and cult around it is critical to its commodity value. This conspiracy of art and marketing becomes thematic in Warhol, but is, we now know, the long result of the formation of the system of modern arts, which is a system where aesthetic values—of genius, innovation, depth of experience, aura, and the like—partly set commodity value, while, within the system, having the permanent liability to reduce to mere marketing formulas or logos.

Warhol made art when Picasso was old—old enough to have become the Picasso phenomenon written about by John Berger in *The Success and Failure of Picasso*.[8] Picasso's long-standing celebrity, the result of his appearance in a thousand newspapers, magazines, art journals, photo shoots, turned Picasso logo as he began to repeat himself, plate after plate, bowl after bowl, posing for the art photographer—bare chested with his piercing eyes, drawing in the sand for the camera, eating raw fish, living in neoclassical eternity in the eyes of millions. Visiting him there was, one supposes, rather like visiting Grace Kelly as he churned out yet another south-of-France azure ceramic bowl or painting of night fishing. You got your two hours of genius on display and went home to publish it in the magazines. The genius of his endless innovation ended up in this assembly line of individual creations and endless poses, each reeking of the name PICASSO, so that the cardiologist or lawyer who read the magazine and then purchased the plate would be assured immediate recognition value for self and friends. This price of the Picasso object was set by the cult around his genius, the aura of the work, but also by the marketing logo of the product. With Picasso (and after) genius took on an aura beyond all commodity value and in so doing set the value of the commodity, the price of the assembly line of art items.

There is only one star icon; she is not repeatable; you cannot have an assembly line of mini-Dianas in the manner of a "mini-me" from *The Spy Who Shagged Me* (1999, directed by Jay Roach). Except in a Warhol picture, and this was his true genius: to realize that the life of the icon is one that celebrates and sustains the aura of the individual and multiplies her into something that has the permanent capacity to reduce her to self-logo, mere consumer item. The life of the contemporary artwork and the life of the icon have the same features. Both exist between aura (the aura of genius, the aura of stardom) and product. It was because Warhol recognized this in his own art and in the art world that he could similarly recognize it in the icon. That was, to my mind, his epiphany, his sense of who he himself was as artist and consumer.

These star icons are new and Warhol is new in the history of aesthetics because they have this life within media and consumer society. Their role is to break out of the consumer chain, generating deeper and more emotional cults, but also to turn back to the consumerist stance. This is new to history, I think. Marie Antoinette was a cult figure yes, but could not yet be a modern *icon*. It took the media in relation to late capitalism to transfigure cult figure into icon, to reestablish the terms of cult as focused on icon celebrities. And it took Jackson Pollock on the cover of *Life*, Picasso's mug shot everywhere, to establish the cult of commodity as a cult maintained by aura and genius, transcendent stardom. It is the joint project of consumer society and the media associated with it to have produced the star icon or artist and his or her public in one fell swoop.

One should therefore not underestimate the stunning force of the art work of genius, or the star, in people's lives—nor underestimate how quickly that turns into product value. Warhol had been stunned: stunned since he, the immigrant working-class closet gay boy sat in his working-class house in Pittsburgh awestruck by Shirley Temple, Ingrid Bergman, Cary Grant. We have felt the same, and also about a Picasso oil painting. We live with our stars like intimate friends and religious saints. Warhol once described his idea of happiness to be going to what was then the lunch counter downstairs at the Waldorf Astoria Hotel, Oscars (in the 1960s) and ordering a soup and sandwich. You choose it, you eat it, it is gone, but it is the same soup you've eaten since childhood, mother's milk, an experience of plenitude only knowable for someone who has eaten and adored the same thing since childhood. Repetition is not only about the conditions of reproducibility, the circulation of the image, the reduction of difference to homogeneity; it is also about an endless supply of things, a sense of their inexhaustibility. Films live forever, and there will always be Campbell's, even if the size of the can has changed. And I don't just mean new films, I mean Marilyn's, which one never ceases to watch in amazement. These are our Campbell's. And Warhol just as quickly realized that when you paint the star who has stunned you in duplicate and triplicate she flattens out into something you could (more or less) find in Bloomingdales.

It is this double life of the star icon that makes her so different from any mere celebrity, this double life of an art work that makes it so different from (more or less) everything else in Bloomingdales. Celebrity is something we know how to understand; it is a system of production with history and economy that, like all markets, expands and contracts until the mode of production falls into

disarray and something else happens. And we understand commodities pretty well.

Icons are different. We do not know how to understand them.
They dazzle and they stun like a stun gun. This aura, I have said,
feeds into every aspect of their lives, including price. Warhol got
it, was sent reeling by their allure, their diamantine power. Behind
his dark glasses the man was permanently glassy-eyed. He'd been
glassy-eyed (starstruck) since childhood. Warhol's pictures are
about their doubleness: on the one hand effervescent, beautiful,
inexhaustible, on the other totally consumable, products among
others in the department store of things bought and sold, circu-
lated and collected. On the one hand amazing, cult forming; on
the other something you can purchase and throw away. His pictures are expressive of both: they reduce icons to consumer products while being icon creating. And they *are* both consumer products and art icons. Philip Fisher, in a fascinating book,[9] argued that
Warhol's work reconciles the industrial and inhuman with the
handmade by celebrating "the factory," while also inevitably carrying the personal stamp of Warhol's "touch" (*fato a mano*). One
might add that they reveal the art in advertising as well as the advertising in art, something a master of advertising art (Warhol in
the late 1950s) turned fine artist (in the early 1960s) would know.
Warhol's quality of individual touch is also, explicitly, a marketing logo, a "product brand," a self-marketing element. There is,
nevertheless, something appealing in Fisher's idea, for Warhol's
work is loved because it is unmistakably his hand, his touch, his
way of doing things, largely unrepeatable by others. His work is
itself iconic. This becomes its marketing logo, but also the thing
about the work which astonishes. In Warhol individuality exists in
a powerful, if also fragile, union. His work, both art and commodity, mirrors the status of the star icon.

And this is, I've tried to suggest, the truly stunning thing about
the icon: that it also seems to unify a double life between aura
and consumable. This is especially bizarre because the icon is neither person nor image but *both*—both in an incomprehensible

collusion. The person's life is folded into the icon, which then folds back into the life, transfiguring and disfiguring it. This is a relationship created of actor and star, which then jumps off screen to become living melodrama. There could be no icons without film stars, which means without the medium of film, because it is there where the audience attraction to the star is born (where the star is born), where the public receives instruction in the way stars are formed of actors, and actors of stars. But that is not enough: TV and talk show and tabloid enter the scene, and the marketplace of consumables, which allow the actor (Grace, Diana) to live lives in parallel: caught between their star quality and its temporality and their ongoing daily affairs. These two never reduce to one but rather braid, like a well-wrought hairdo, a wig or a noose, depending.

This braiding fulfills a deep desire of Romanticism: that life and art should become one, life thereby exalted, art thereby perpetual in life. This is Mozart's dream of Don Giovanni, Giovanni's dream, the dream of the *Lebenskunstler*, the artist who lives life like it is a work of art, making art of life. The sense of power, of omnipotence makes this a male fantasy. There is no room for the gap between desire and fulfillment to arise, nor place where imperfection might triumph. But if the dream of Romanticism is that of a life of complete power, both beautiful and sublime, of complete union (between person and art), the price of this braidedness for the icon is nearly complete powerlessness and to be caught in the abyss between her two lives. She is a passive entity in the formation of her persona: plagued by the halo around her self. That halo is the creation of the media and the audience desire channeled through it. Public desire hangs onto it, and her, forcing her into the position of recipient (if not bug under glass). She cannot exit, except by rushing off to Europe or slamming into the concrete side of a Paris tunnel. Her life is braided into an art she has in no way made, even if she has participated in its making, which she in no way controls. With the icon, the Romantic dream of power is transferred to the *audience*, which gets to create and enjoy her double life (between person and art/icon) and this is the vast aesthetic work of the media. Without a long look at that media, we will not be able to understand how the process works, nor appreciate its terror. With the icon the terms of power reverse. It is the public, rendered powerless in its mere voyeurism, that ends up controlling the icon—and through its own projective desire. The icon, who is given the power of the silver spoon, ends up caught in the spider's web of the silver screen.

But there resides an irony: The icon figure is powerless before our gaze, but we, we are also powerless in our awe before it/her. As we shall see later, this double position of power and powerlessness is what film theory has understood to be the position of the film viewer watching the star on screen. It is a position Alfred Hitchcock revels in exploiting, a position his films are inevitably about. I am fascinated by a world in which everything powerful is in fact without power, a world of relinquishment. It is a dangerous, religious world, a world very close to idolatry. I am fascinated because I am a Jew grown up on the absolute rejection of idolatry. Thou shalt have no other gods before thee, only the one who shall not be made visible in the form of an icon. In our contemporary world everything seeks to be made visible, and visibility conveys price and power. And mutual powerlessness, because Diana's viewers kneel before her and are powerless to change her life, while she can hardly get away.

The icon is as close to an idol as one can get.

three

Therefore Not All Idols Are American

THE ICON is as close to an idol as one can get. Diana was an icon. Therefore not all idols are American. Diana was loved in a specifically British way: for being the royal who broke the mold and became the "People's Princess." She was a being who brought the aura of royalty to the people and did it by baring her pain—and joy—to the public and by breaking out of the cast-iron role of Buckingham royal. She did it by seeming (and being) human, rather than scripted, by identifying with humanity in a way out of sync with the royal position of statuesque. That her pain and joy were the stuff for millions meant she was anything but common—a celebrity at worst, a film star at best—for ordinary people are not watched by millions, who would quickly lose interest in them. Nevertheless, Diana refused the British stiff upper lip, and the purdah of the British princess, to walk among her congregants and speak her emotions (that is, emote) in a language they understood to be on a plane with their own. This was a major breakthrough in the domain of English reserve, a failure of tact so monumental that Prince Philip was ready to send her to the dogs. Her public could revel in the fantasy that, through her, royal and commoner became one, while remaining awestruck that a royal could deign to walk among them. This reflected badly on the British royalty, whose refusal to face the media and disgust with her media role led to a huge public backlash.

"At the end of the twentieth century," King Farouk was said to have remarked, "There will be five queens left, four in a pack of cards and one British." The British monarchy is, apart from the Japanese, the only real game in town. London remains the only world-class, cosmopolitan city largely owned by two royals, the Duke of Bedford and the Duke of Marlborough. Owning a flat, however posh, means renting the use of it on extended, hundred-year lease from one of these characters. A stockbroker with a seven-bedroom flat worth millions is as often as not one of their "vassals."

British royalty remains a matter of lineage and is treated as such. The Earl of Spencer did not miss a beat in reminding those listening to his oration that the Spencer family is of an older title than those comparative upstarts, the Windsors, and that it is because of Diana's pedigree that her children are as thoroughbred as they are. Whatever else was and is true of Diana the icon, she was and remains a historically conditioned icon. Her aura as film star could only have been as it was and is because of the longevity of her title and distance from the populace royalty presumes and assures. Title lends entitlement, which is something Americans wholly fail to understand except in the abstract. And it also lends the desire for intimacy. "Her majesty's a pretty nice girl, but she doesn't have a lot to say. . . . I want to tell her that I love her a lot, but I've got to get a bellyful of wine. Her majesty's a pretty nice girl. Someday I'm going to make her mine, oh yeah" (*White Album*, Abbey Road Studios, 1968). This Beatles ditty is pure working-class Liverpool; a plumber or housewife could have made it up while temping, taking out the trash, waiting for a bus. It could have been sung in a corner pub after the third pint of bitters in male chorus. The joke of intimacy with a woman who hardly appeared on the telly, much less in a home or even local church, whose charitable works were pure formalism and whose language was scripted in an archaic English where the letter "r" comes out "w," is quintessentially English. And no queen has been more awkward with her people than Elizabeth II. Helen Mirren masterfully acts her awkward arrogance in *Queen*, where "Mam's" only spontaneous warmth is toward her dogs and a lone stag being "stalked," itself an irony given that the story revolves around Diana the huntress/hunted. (The queen can cry for a hunted stag, beautiful like Diana, while resolutely hunting her own, ex-daughter-in-law.)

The fantasy of intimacy was, however, given new dimension with Diana, rejuvenated, since her life was in rebellion to the crown she wore and her instinctive energies were directed toward "the people." This was a woman with whom the public identified and who identified with "her public," if in the stilted and dramatized terms of the royal. Hers was a chapter in the historical story of the monarchy, in the story of monarch and people, in the story of a monarchy at odds with its own commoners and, through her, despised.

The grace the British public vested in Diana drew on a long tradition (since Henry VIII and his break with the Catholic Church) tying religion to royalty. The Church of England is just that: Church of *England*. The reigning monarch has the official title of supreme governor of that church, even if the senior cleric is the Archbishop of Canterbury. In Westminster Abbey and the private chapels of the entitled, class converges with sacrament. Diana's crown carried the intimation of high Anglican grace, but her people's princess rebelliousness drew her to those outside the top rung of the religious ladder. Through her the public could find a royal route to grace and a commoner's longing for it. Diana, in fleeing the life of royal, proved in need of a grace more human and rewarding than Anglican mercy. With Diana, grace and disgrace became a single saga of woe, and the religion of the state (with the public awe it engenders) converged with the human melodrama, placing her in need of religion.

There is no equivalent in American life for the historically lineated (and graced) royal, nor for the state character of religion and the way it dispenses awe on the basis of class. America is too much in love with its commoners, too much in enmity of king and crown. This leaves an open space for the hero to emerge, the film star, the high-volume gangster and T-shirt Brando, the Kennedy male. It leaves a gap in the logic of grace. The Kennedy male was (and is) a rough-and-tumble Irish gangster type who wanted ass with class, which for Joe, bootlegger turned first chairman of the Securities and Exchange Commission under Franklin Delano Roosevelt, meant silent film star Gloria Swanson. Jack migrated up the pecking order with Jackie, became war hero, which raised him above his Irish/gangster origins while highlighting their male-can-do character into a profile of courage. He wed Miss Upper-crust fashion queen while maintaining his father's lifestyle, which meant Marilyn. Jackie was so fed up with Marilyn in the White House, and with the compounding headaches of the Kennedy compound, that she fled to Europe ASAP after his death and stayed to raise her children away from all that. Their Camelot existence was an American dynamic of man of the people rising to the heights of the American century with a six-gun, a film star, a Congressional Medal of Honor, a wife dressed by Oleg Cassini (who was, to complete the cycle, for a time Grace Kelly's special friend). It was the American fairy tale of a world where Europe was struggling to emerge from its ruins, Russia was the spy who loved me, and the dollar was king. Might made right and was right, when Kennedy exemplified it. This was all the lineage the unruly new empire required. The American hero is a man of ideals, his vision comes from the commoner, his instincts remain macho in the form of gangster and western films. Where Britain seeks pedigree America seeks cash and carry and Hollywood value: the Kennedys were fabulously rich, and by a crime (bootlegging) that rang mythical (in the name of Jimmy Cagney). Get enough money fast enough, build yourself a compound, sleep with film stars and aim for the presidency, and you will have your dominion. This is all the inheritance you need. Add melodrama, T-shirt good looks, and charm to beat the band, and you're an icon. These boots are made for walkin', that's just what they'll do, one of these days these boots are gonna walk all over you. Nancy Sinatra strides onto the Ed Sullivan show in white, glaring, patent leather hugging her legs all the way to the insides of her thighs. She has a microphone in one hand and lip-syncs like a frozen Hitchcock blonde. She's ready to trounce everything in her path, and the boots announce it. Oh, that Virgilian tune, it's so elegant, so intelligent! Behind Nancy lurks Frankie, drinking his guts out in some Vegas bar, his Vegas nerve ready to explode if his daughter cries mistreatment. He'll send a few capos to break your legs; that's what they say. He's drinking bourbon with the rat pack, Dino, Sammy Davis, Peter Lawford, spending greenbacks like water and sharing women like bottles of gin. Behind them, glowing like an orb, are the Kennedys. Their stardom found its confirmation in this firmament. America on the march, the power of their feet, an adoration of John Kennedy and John Wayne, John Glen and Jonny Guitar. American icons are posthistorical in that mythohistory is replaced by immigration and rags-to-riches Parsifal history, a history of immigrant rising to the occasion so fast (one generation, at most) that it seems instantaneous, as if it were cinema. It's *Citizen Kane* stuff, America on the march, larger than life stuff, thirty-five millimeter. And where you can rise up you can equally rise down, swooping with the braggadocio of George W. Bush, pit bull who grew up with a silver spoon in his "purebred" Connecticut, Kennebunkport mouth, slumming himself down through the pretense of "bein' from Texas" and "speakin' like a guy who just walked out of the local bar" and, with oil money gushing out of his every orifice, turning this elective affinity with white trash life into a presidential career. It would be literally unthinkable that British royalty be in such a self-identifying slide *down* the class ladder. Which is why all mention of Dodi Fayed was barred from the funeral: he was Muslim and worse, as common as they come, merely being from cash. He might as well have been from Texas. Fayed would by contrast play well in America except for the Middle Eastern bit, since Fayed's cash would confer the equivalent of what is in England historical status, and his commonness "old boy authenticity, the kind a red-blooded American can believe in." It would not make him a star but certainly a celebrity, and if he had that special something, he might end up a star. Diana was best in show, the purest of the pure (and lily white too); Kennedy has gun running behind him and a wad of dollars and Jackie and Marilyn and Oleg Cassini and the CIA.

In America the film star is as close to a royal as one can get, since the days of the decline of the Philadelphia elite anyway. This is why being from Philadelphia, and well bred, played a role in Princess Grace's persona only through her well-formed, well-bred demeanor in films, not through direct historical lineage. What catapulted her into the firmament was her ebullient, seductive, perfectly blonde and impeccably mannered physiognomy as actress, along with her upper-class disdain for a life of it. She took up with the Oleg Cassini who had dressed Jackie and then became the perfectly bred princess of the ersatz kingdom of the five-card stud. Breeding yes, lineage no: hence the "ersatz." A Grace Kelly would never have made it to the British throne, and not only because her last name is Irish. Remember the fate of Wallace Simpson and Edward VII.

Diana carried a historical pedigree that, turned into aura through its social force and the habits of monarchy and filtered through her physiognomic beauty, was already unique in the world. Her iconic value was predicated on lineage, but broke through that into film star quality, on the one hand, and TV soap and tabloid, on the other (the people's princess, woman on the verge). In the Diana phenomenon the posthistorical and the historical fit together like a glove. She really was a figure of reconciliation: between the historic value of a royal that is vested by lineage and the posthistoric value of a celebrity that is vested by the media. Except that she nearly broke apart the monarchy in being what she was, not to mention herself. Still, the difference between the British and the American icons, with Diana and the Kennedys in mind, is that the British derives from historical lineage, the American from money and rags to riches mythology. With Diana, the public transposed the cult of royalty to the cult of the media star. With Camelot, they transposed the cult of the Horatio Alger gangster and its rise to the cult of politics for a charmed nation-state. The Kennedys were the first and last royals in American public life. Diana was the last royal and the first media star in British public life. Because of her, the royalty cracked. At the end of *Queen* Elizabeth holds out the white flag to the media by appearing on television and speaking about her hated ex-daughter-in-law. She knows the continuation of the royalty demands its acknowledgment that the media has won.

There is one feature Dodi shared with the Kennedy clan: he was the son of a tough nouveau riche on the way up. Rumor has it (which shall forever remain mere rumor) that he proposed to her, or planned to, that fateful night at the Ritz. That she would have even approached this level of intimacy with a Dodi Fayed is a final subversion of the monarchy, of the very class system itself. She was truly the people's princess, even if finally on the way to becoming a Grace Kelly, princess of an ersatz kingdom run by Middle Eastern nouveaux riches, the kingdom of Harrods, production company and tax shelter.

four

A Star Is Born

AESTHETICS, ESPECIALLY philosophical aesthetics, has tended to approach its topic singularly, zeroing in on this or that art as if it existed in isolation from others. And yet the modern system of the arts forms just that: a system. Increasingly today the forms of aesthetic appreciation cut across individual media, and are the product of many in particular consort. The world no longer comes pure in the way the abstract expressionists or "absolute musicians" wanted it, but as a series of hybrids. Nowhere is this more true than for the cult icon, and it is time to turn to the systems that produced Diana: the star system, originating with the tabloid, then the media of film and television, and finally the nature of consumer society. (Royalty has already been discussed.) The challenge is to understand how these differentially create and manipulate the star icon. In pursuing this systemic approach (which I believe new in aesthetics), I shall want to allow the media to speak about itself: especially film, which at its best explores its own aesthetics and social position in ways from which philosophy has much to learn.

I shall begin with the star system, which is a product of the nineteenth century but vastly expands when tabloids combine with film culture and then again when television enters the scene. Here I shall rely on excellent work already in circulation. For this system has been well explored.

David Gritten tells us that the star system came into being "around 1900 . . . [when] moving pictures became available to the public, and a genuinely popular press emerged. What united these events was that both industries in their early days grasped the need to invent celebrities to advance their expansion"[1]. Gritten goes on:

> Alfred Charles Harmsworth . . . founded Britain's first popular news-paper, the *Daily Mail*, in 1896. Sensing huge profits could be made in entertaining readers rather than providing a public service, he set himself against a tradition of papers with long, dull stories in long, dull columns. Aiming at the middle classes, the *Mail* reflected in its pages the topics its readers might discuss or gossip about in their everyday lives. He downplayed reports about matters of state, for-eign policy and economic questions and instead gave prominence to sensationalist accounts of crime—murders in juicy detail, confession letters printed in full, and the like.[2]

In America Joseph Pulitzer and William Randolph Hearst car-ried this turning of news into saleable entertainment further. When Hearst took over the *New York Journal* he early on published his philosophy:

> "The public is even more fond of entertainment than it is of infor-mation." He famously invented the war against Spain in 1989, and "encouraged journalists to insert themselves into their own reports, sharing with readers the ways in which they virtuously revealed injustice and wrongdoing." In doing this, of course, his journalists embodied entertainment values—creating stories in which they had the starring roles.[3]

Joshua Gamson puts it this way:

> As the newspaper industry took off, publishers fighting for a competi-tive edge in the increasingly information-dense environment, spurred by the circulation wars and the introduction of "yellow journalism" in the last quarter of the nineteenth century, made stories about *people* a central feature of journalism. In particular, newspapermen like Joseph Pulitzer and William Randolph Hearst sought "human symbols whose terror, anguish, or sudden good fortune, whatever, seemed to dramati-

cally summarize some local event or social problem or social tragedy." Names, in short, began to make news.[4]

Hearst turned the journalist into self-promoting celebrity, rather like Stanley, sent by his newspaper to find Livingstone, "lost" in deep, dark Africa: sent so that he might write about it from the personal angle and break headlines. What he brought back were the immortal words: "Doctor Livingstone I presume," which shall live on long after the newspaper he worked for has been forgotten, roughly in the way that a generation later Chaplin will live on even after the public forgets which films he was in.

At stake in the turning of journalist into celebrity figure was expanding market share. If readers could be kept hanging by a thread because the story was so good, or develop loyalty—if not fixation—upon particular journalists, then circulation would expand. Newspapers were central in the creation of the modern celebrity, by keeping his or her story alive and circulating it to the hordes of hungry readers. By focusing on personality as a central ingredient of story, they turned the persons themselves into the substance of the story and the reason for reading it. By turning the journalist into a celebrity, they kept interest in the newspaper growing.

Once the movies came into existence a new form of celebrity was created, the film star. This was the joint work of the movies themselves, seen by millions in the full glory of their qualities, and of the studios, with their ingenious forms of build-up. The early movie Mogul Carl Laemmle, for example, introduced a new technique for generating public interest: that of the disinformation campaign. As Gamson says, "he hired Florence Lawrence, whose face was already recognizable as 'the Biograph Girl,' and apparently planted a story of her tragic, untimely death. In Barnumesque style, he subsequently denounced the story as a lie and as proof announced Lawrence's appearance in St. Louis, where she made a tremendous publicity splash."[5] Gamson continues:

Movie manufacturers, now firmly committed to mass production, adapted the star system to the industry's needs. In particular, producers needed product differentiation to stabilize demand and price, especially crucial given the intangibility of the film product. Early attempts at demarcating their products without using players had not been especially successful. At first, producers used their own trade-marks, which apparently were not distinguishable enough to fans. Next, they tried narratives. In 1911, Motion Picture Story, an early fan magazine, asked its readers to choose their favorite film stories, and [the magazine was overwhelmed with demand for information about players]. The fact that audiences distinguished films by stars became inescapable, and accordingly, the producers began to utilize talent as a successful strategy for differentiation. . . . The advantages of the star system had become abundantly clear to film manufacturers, and the studios moved quickly to institutionalize it.[6]

Thus was star born and, in the early days of studio management, built up as ethereal, while also pushed as "ordinary." Nothing was more American than this. On the one hand the studio kept the star at a distance, on the other encouraged personal intimacy with her. She was a distant planet, worthy of awe, but also put forward as a person of talent and integrity, as if these human qualities lent her natural born right to stardom. This was also true of the British Ealing studio star, who was a figure of refinement as well as ordi-nary good will, often a secretary or middle-class housewife. The female star's exhibiting goodness was quite different from the Wei-mar film, where the star was anything but (a Louise Brooks eating men alive). Gamson suggests the American studio's need to make their stars out to be good, ordinary people, off screen as well as on, was its way of encouraging belief that the star was not a manu-factured entity—no artifact of studio hype but rather natural and well meaning, a hometown girl whose stardom was as self-made as any Horatio Alger's of the myth. This propagated falsity helped maintain public credibility in a media that might otherwise have been accused of massive manipulation and moral indecency from the teetotaling set, the American farm, and the small town. Even the men had to be good, however much they were macho cigarette types from the wrong side of the tracks. Any hint of public scandal like that of Fatty Arbuckle and their careers went off the rails. The star is a rare commodity, and the paradox of the studio system was to establish her rarity, her exclusivity, while also marketing her as the girl next-door. Diamantine in her inaccessibility, she also had to appear as salt of the earth, mere carbon on the job. Mary Pickford was thought never to have lost hometown values even as she rose into the firmament. And so the censor board, the small town, and the American populist might, in these early days of film, rest easy with the system that was making big bucks off the star system. And so the public might rest easy (in these days of moral and legal prohibition) with the star.

The image of the good little boy or girl turned star in virtue of their talent and goodness, the goddess still at home on the farm or in the local penny candy store, allowed America to misrecognize itself as the simple, hometown place of Grovers Corners and Bedford Falls, the Tom Mix wild west of good guys trumping bad. These good guys were those who trumped the Indians. The west, already colonized by 1900, lived a second life in the wild through the Western, whose birth was coterminous with the end of the westward-ho expansion. There were no more wagon trains except in movies where they could become living mirrors of what Americans wanted to see about their past and present. And so a wild west of good citizens, protected by the six-shooter and in love with their own struggling communities, and so the good girl in the bad city, an also wild and unruly place, the girl who is preyed upon by urban Comanche warriors, gangsters, and all manner of tough guys while retaining her Mary Pickford innocent eyes and white on white kitchen apron. The studios played from the beginning a crucial role in creating the ideology of nostalgia by making it so absorbing. The star was the immigrant writ large, the children of my grandmother who, as she drummed into their heads, were destined to rise in life to the top of the Empire State on account of her traipsing the streets of Salem, Massachusetts with arthritic fingers that refused to stop sewing, her implacable will, boundless energy, and the fact that life, after two millennia in the Diaspora, owed her big time.

This will to perpetual rise in a state of innocent grace can be found in that great artifact of American optimism, the MGM musical. In *A Star Is Born* (1954, directed by George Cukor) Judy Garland reveals the power of the chanteuse, the glory of the dancer, but also an attitude of Midwest gumption and old home decency that her public adored. In the first half of the film, the Judy Garland character portrays her own rise to fame in a musical sequence about a young girl arriving from the hinterlands to New York, banging on the doors of agents and producers, showing her wares (dancing and singing). She is rejected, misused, mistreated by unscrupulous men, while always remaining pure and decent, never slinking onto the casting couch, never failing in vision, always sticking to it until finally she gets her chance and wins the prize: stardom and success. It is the immigrant arriving at Ellis Island all over again, the young girl winding her way through the dirty streets of lower New York in search of work, a single silver dollar in her shredded coat pocket, struggling in the sweat shops, sewing into the night by the light of a single candle, until her fingers are swelled and arthritic, so her progeny may inherit the destiny she is utterly certain that, like God's portion, is her entitlement, until the next generation bursts onto the scene of splendid postwar wealth, at which point it's all Roth's *Goodbye Columbus*: the house in Westchester, golf and bikinis at the country club, plush leather seats in the automobile, a maid called Carlotta to scream at between the chicken and the fruit compote because she didn't put ice in your drink. It is Iowa farmer (another source of silent film melodrama), slaving day and night through blizzard and tornado, struggling against unscrupulous banks and loan sharks, always in the front pew at church, his beloved girl preyed on by itinerant actors and dashing knaves, until he gets his chance at a new tractor and for the first time saves money for his son's college. The star system is rooted through Americana and its self-image.

A Star Is Born is also about the star system insofar as it lambastes the crew of professionals that arose around it—the agent, the producer, the casting director—those Jack Carson types who will endure all manner of insult from you while you're up, then screw you when you're down. These are, along with psychiatrists and German war criminals, among the characters Hollywood most loves to hate. These are the ones who expose weakness and generate anxiety. Before the agent everyone is afraid; before the psychiatrist castration is feared, or manipulation, before the Nazi, annihilation. Along with snakes, zombies, Frankensteins, and denizens of the South Bronx, these are the characters you don't want to meet except when protected by Clint Eastwood or John Wayne. They point to the difficulty within the system: that it runs on cooperation from people who carry knives in their back pockets, that its system of trust is a sham because it is about putting over a fast one and making a fast buck. Within this system even the most successful can be wiped out in an instant, washed up in the pool of their Beverly Hills mansions with only their dog as a friend and no one answering their phone calls. Witness Fatty Arbuckle, Erich Von Stroheim. Even the greatest of them, Orson Welles, ended his life in a wheelchair for the overweight, doing Gallo wine commercials to pay for his supper. Recently Mel Gibson, arrested for speeding, failed an alcohol test and caused a scene, accusing the Malibu cops of being rabid Jews, screaming that it is the Yids who start all the wars in this world (this at a moment when Hezbollah had just attacked Israel). The scene made the news, especially given that his father is a known Holocaust denier and that his film about Christ has been accused of anti-Semitic overtones. Within days his planned series on the Holocaust (but why is he making it?) is canceled by the American Broadcasting Corporation, and Gibson, obviously told that his career, however powerful in the past, will be in jeopardy if he doesn't go before the media and eat humble pie, appeals to the Jewish community that he is not an anti-Semite, apologizes for the incident (although not for his remark about Jews causing all the wars in this world), and asks to meet with Jewish leaders, "with whom I can have a one-on-one discussion to discern the appropriate path for healing."[7] "I am in the process of understanding where those vicious words came from during that drunken display, and I am asking the Jewish community, who I have personally offended, to help me on my journey through recovery" he weeps in

Crocodile Dundee tears.[8] Only in a culture in which Mr. Braveheart may rehabilitate himself through the language of sworn victimhood, as in "I'm an alcoholic buddy and I needs me a twelve step program to keep me from talking like that and, moreover, it was the drink talking, not me," or "yes I'm an anti-Semite but it's a disease, like alcoholism, and I should be treated with the respect accorded a cancer victim." Gibson's response was a mixture of public confessional and public excuse, something only a culture drenched in talk radio and TV could pull off. Later he goes on the Diane Sawyer program and says he was sore because it was the Jews who had lambasted his movie about Christ, as if that justifies everything.

The Gibson affair shows just how gendered the role of star persona is and has been within the star system. Mr. Scottish Warlord, with sword and iron club, became momentarily feminized in his display of confession. "I was out of control, unable to contain myself, weak, badly displayed, I am ashamed, I feel terrible, forgive me, please don't drop me, etc." This is woman stuff—the stuff of woman's TV (soaps, talk shows, etc.). He hopes it will get him back in the male groove, but in order to get back there he had to become a woman. The moral: one false move and the studio drops you, you can't pay your psychiatrist bills, and you lose your house to an ex-Mafia boss who is now thriving in a second career as a Hollywood agent with the help of the FBI witness protection program, and the only way you have to get back is to act the crying female. This is not Diana land, not the place where stiff upper lip goes down well. No, you've got to bare your breast and hope the silicone implants do not leak. And so it goes, up and down the ladder of success with the craziness of a Jerry Lewis comedy.

More than that, the Gibson affair shows how creative the media are in absolving themselves of their own corruption, displacing it onto the bad guys of their films, baptizing themselves in the waters of the talk show confessional, whatever. In the 1980s everything changed and the American public fell in love with the bad guy, who is now the subject of such HBO serials as *Entourage*, with its cast of underage brats and agents who would kill their own grandmothers for a second Porche. But celebration of media corrupt-ibility cannot go too far either. Too much and the public still balks because no one wants to believe they are truly a sucker. Nor that their society runs on dollars laced with traces of cocaine and grimy fingers. America is in love with the cash that trumps all, the celeb-rity with the toupee that is Donald Trump, the opera with Tony in the soprano role. But it is also in love with the clinging nostalgia of its Grovers Corners goodness, its moral right to stardom. Put those ideals together and you have a system that constantly veers from one side to the other.

The next component in the aesthetic genealogy of the star is the cinema itself: the aura created therein, through which the star can become a being inhabiting a distant planet. Without the aura, and the movies that produce it, there could be no drama about Judy Garland's goodness. Not to mention no Marilyn, Jackie, Diana. But what is this aura that comes from the movies? How can it be best understood?

five

The Film Aura

AN INTERMEDIATE CASE

WHAT IS this star aura that movies seem to radiate? There are celebrated theories that deny film has an aura at all. The issue is critical for understanding the aesthetics of the star icon, since, whatever else is true of her, she casts—or is at any rate perceived to cast—a glow. The star icon exists in a halo, which seems to derive from her media presence. The question for aesthetics is how best to characterize it.

What is an aura? An aura is an atmosphere or mood, a halo surrounding something—a ruin, a pile of rocks, a site of battle. It may be the halo around a holy person or princess. An aura is not exactly a *property* of the thing or person in question—not in the way size, shape, materials, or color—even consciousness—are properties, anyway. Rather, an aura emerges through a perceptual transaction between the properties of the thing/person and someone viewing it or taking it in. An aura is a way of seeing, hearing, feeling a thing, an intersubjective image of it, a way of imagining and responding to a thing. Things or persons are invested with auras through direct experience of them and also through representations that lend them special signs or transfiguring moods. We come to be trained to see saints as having auras through their Fra Angelico representations with glowing, childlike halos. Through the experience of the representational image, devotion is inculcated. In a culture of devotion image can become icon, and icon can incant aura.[1] Narrative is central to the formation of such devotion. Knowledge of story is what often lends the thing its special atmosphere. The stones of Stonehenge, the source of the Ganges, the Elvis mansion, the cathedral at Amiens: these are all sites of worship because of larger stories and myths around them. Everything we bought in the 1960s had an aura, from our psychedelic album covers to our posters of rock stars to the weed we smoked, because life was for a moment turned into a cult (until the oil crisis hit in 1973 and we all got deprogrammed). The sixties were a critical moment in the marketing of the aura, which is part of the larger story of aura in consumer society.

The locus classicus of twentieth-century writing on the aesthetics of the aura is the work of Walter Benjamin, specifically his essay, "The Work of Art in an Age of Mechanical Reproducibility,"[2] where Benjamin points out that the aura has its origin in religious cult practice. For a thing to be viewed as having an aura the viewer must stand before it, be in its presence. "Be here now" was, not incidentally, the slogan of the sixties. The young Moses stands before the burning bush in the starkness of the desert, alone, to witness the power of God (God's aura); the Druid dances at Stonehenge as the dawn breaks because the stones carry a spiritual magic to which one must be present. The Russian saint worships before a particular church icon, meant to be the container of special properties.

There is a shift in the quality of aura from cult magic to aesthetic form as religion gives way to secularization in the history of a society. Stonehenge to the ancient Druid is a place of power, commanding awe. Its value is sacred, its role, ritualistic. This changes utterly as Druid culture is lost but the thing remains, an empty signifier of forgotten powers, site of strangely hovering magic to us and speaks like a screen memory of something profound that we can no longer capture. This lost sacredness, now returned as hovering and oracular, is what Benjamin meant by the aura. Stonehenge once had a power. When that power evaporates the place is left with a new kind of aura, carrying the unspoken intimation of the past. The aura emerges as an aesthetic phenomenon of modern times only when original ritual values attenuate or pass away. It is a secular form, a rune of lost magic of a culture no longer "there." This power of signification is the birth of a new experience, that of faraway powers and lost times concealed in the presence of the thing. Art takes over from religion as the form of the sacred, investing the world with the power of this new aura. The aura is the presence of the long ago and faraway. It is a script of shadows, a living cinema of displaced images before which we stand.

The secularization of ritual value into aesthetic practice is by the nineteenth century sufficiently complete that cult worship is transposed from church to museum, concert hall, and grand tour. The museum, that bastion of the eighteenth century in Europe, is already a perfect place for the formation of aura, since it has removed objects from their original sites and turned them into sights, things arranged in abstraction from use, worship, circulation, cult, things ready to live second lives in the imagination of the museumgoer, for whom their masklike presences carry the implicature of hidden, unknown powers. The birth of aesthetics, coterminous in the eighteenth century with the birth of the museum, is predicated on the abstraction of object from site. For only then do objects become masks, hiding soul, context, circulation from the viewer, carrying feather, jagged form, monumental size as a secret sign of lost place, lost soul, licensing a power in the viewer to reinvent and interpret the aura at will. Orientalism arises in part from this license and the projection of monstrosity—but also the pleasure in an aura that remains forever hovering in the air, since the object is not returned to the site where it once had a genuine life (like Stonehenge).

When Europe has acquired sufficient patina, when its ancient stones, venerable ruins, secret caves and rivers, glorious cathedrals become original and unique things, known for the long history they carry, Europe becomes the bearer of the aura. And so Franz Liszt can spend the season at the Villa D'Este and write a volume of his piano music, *L'Annes des Pelerinages* (The years of pilgrimage), neither about Jerusalem nor any crusade, but instead in praise of the iridescent, September light playing over stone and garden at the villa itself. When pilgrimage is the grand tour, Goethe down the Brenta, later Edith Wharton and Henry James in Rome, the transfer of religious sentiment into the worship of object and place is complete. "You must watch the sun rise over the Veneto from the furthest island of Torcello, and then watch it set over the Grand Canal from the Island of San Michele. Nothing less will do," the mentor coaches his incipient traveler. "And read Goethe before you take the boat rise down the Brenta to Venice, so you may see it the way he first saw it a century earlier." The young traveler is a neophyte, awaiting instruction from his betters into the virtuosity of special experiences. The neophyte shall learn connoisseurship, be trained in the treasures of the beaten path and how to stray from it, in the art of personal encounters with "Italians," in the charm of foods in decaying locations, in the craft of diarizing and sending home letters by post, even in how to study Baedeker's and avoid getting hit by rushing carriages while walking. These preparations will pay off when the traveler sets foot on the soil of the Continent, gaping at Roman ruins, getting lost in the mazelike streets of the Istanbul bazaar, breathing in the aroma of fig and tomato in the Roman market, and then, one fortuitous afternoon, when the sun is so hot he feels he must flee to Scandinavia this very moment or wilt, he will arrive by chance through the streets of Rome at the surprise of the Trevi Fountain, splendid in its size and spume, a theater of carved figures and dancing water bursting the bounds of a site far too small for it, a grand opera tucked away in a tiny piazza. Were the traveler there at two on a summer morning under a full moon, he would see Marcello Mastroianni kissing Anita Ekberg.

The cult of the aesthete may be sourced to figures of the late eighteenth century such as Baudelaire's dandy, that stern follower of a monastic order devoted to the temple of high art who approaches aesthetic experience with the rigor of karate (and the same level of masculinity). The nineteenth-century traveler is a member of a similar aesthetic priesthood and, even more, is the artist as path breaker in this cult: Gauguin in Tahiti, Mendelssohn at Fingal's Cave, Liszt everywhere and anywhere with someone rich (and beautiful and upper class—if not upper class then at least a soprano), Wagner wearing silk underwear purchased with his lover's husband's money, composing operas in the pneumonic *piano nobile* of a Venetian palazzo while avoiding creditors and talking directly to God. These geniuses produced the poetry of the cult, were its currency. Nietzsche is appalled by Wagner's Bayreuth because it is, in Nietzsche's understanding, a temple of idols where Wagner is Mr. Rajneesh himself with his fleet of automobiles and his legions of *Untermenschen* (not to mention young women).

The Europe of the aura is a Europe of endlessly varied presence (the light as it strikes the stones of the cathedral) and pastness (the patina of the ruin, the ancient character of the village). This is a particular kind of experience of the aura that is different from active cult worship (of the special powers found at the source of the Ganges). The pilgrim finds aura at the site of pilgrimage only if that site is the site of pilgrimage past: the death of a saint, the performance of a miracle, a form of life that was (and if it remains, remains only in the form of a historical addendum).

Diana's aura comes from the long ago and the faraway: it is that of the royal lineage, the secret palace chamber, the special, rarefied experience of life. This aspect of her aura combines with that of something imported from another kind of faraway: film itself. She is a living presence carrying these depths of the past and of the silver screen, just as Jackie is a living presence carrying the image of Camelot, the sense of faraway designer perfection, and then the time-stilled assassination that followed her around like a weight halo. Film stars like Marilyn carry the aura of their on-screen presence into their offscreen lives, and nothing is more mysterious than this transition. But what then is this on-screen aura that a Marilyn can carry offscreen? What is this halo that can attach itself to a Diana? How does film generate the kind of aura Benjamin describes, an aura of presence and pastness before which one stands? How can the cinema screen become an occasion for the right imaginative depth to create aura?

Benjamin himself believed film does not and cannot carry this power. His theory, widely believed, is predicated on this. Writing in the early part of the twentieth century, Banjamin argued that new media like film were in the ascendancy that would completely reshape public experience and consciousness, producing a gap between the Romantic/religious world of the nineteenth century and his own. Mechanisms of mass reproducibility predated the twentieth century, of course, but with the advent of film they acquired the technological momentum necessary to make their true aesthetic effects known. And so now, at the time of his writing in the early twentieth century, the media of photography, film, and printing were making it impossible for the aura to remain any longer in the central experiential role it occupied for the Romantics. The aura was soon to become a thing of the past, like cult worship before it. No longer would the aesthetics of the grand tour and its cult of originals, each carrying traces of the past while exuding peculiar beauties of presence, remain the dominant European form. Destined to wither, they would be replaced by something truly revolutionary. And film would play the crucial role in bringing about this brave new world of aesthetic experiences.

Benjamin never quite tells us what this brave new experience would be. His work ends in prophecy that it will happen, reveling in its vast immensity. Instead he tells us why the old forms would wither, what will no longer be central after the aesthetic revolution. It is the why that is most important to me, for it hits the nub of the issue of film, stardom, and the aura.

Film and other mechanisms of reproducibility, according to Benjamin, are destined to make the experience of the aura wither, because they remove the viewer from direct visual contact with the thing viewed. In so doing these media disallow the thing from exerting its "magic" of presence (through which the past is conveyed). Venice at sunrise is overwhelming, Venice in four-by-four postcards, photos, home movies, films a less exalted thing. The first Venice presents its magic of light upon stone, the reproductions do not. In mechanisms of reproducibility, said Benjamin, and above all in film, there can be no aura because nothing is actually there onto which viewers could find and project a halo. Film (to improvise) is nothing more than light projected onto a two-dimensional surface. Theater, according to Benjamin's argument, does retain the aesthetics of the aura because in the theater the audience remains *face en face* with real, live actors who might thereby exert their magic.[3] The so-called aura surrounding film is, Benjamin concluded (with admirable prescience about things to come) merely a studio buildup of stars—in magazines, two reelers, newspapers, production photos, not anything internal to the film medium itself. No star aura is found within the aesthetics of the medium itself, since the medium lacks a position from which we may stand directly before anything or anyone "in" it.

The result of this is profound. Since there is no "original" in the case of media of reproducibility, before which the viewer can stand, the cult of the original will wither—and with it the grand tour and the elitist aesthetics associated with it. A medium (film) that generates an indefinite number of prints, all equal, is a medium that removes the viewer from proximity to particular cult objects, and produces equality throughout mass experiences. No matter where you watch that movie, in Calcutta, Paris, or San Jose, California, you are as close to it as anyone else. No longer are aesthetic experiences the special and unique purview of the few who can afford them, no longer are the masses left out. Circulation throughout mass culture will desacralize the cult of special individualities. No longer can there be a church of special objects in which each is a crucifix and each person experiencing them a high priest. The aesthetic cult was in its very nature a closed system, a special society dependent upon class and training. The twentieth century would break the circle and democratize art. This is what the twentieth century architect was aiming for when he modularized cities into large sectors, each with sameness of construction and spatial disposition. He was aiming for equality of living conditions and a growing recognition that we are all the same. He was aiming, through sameness of dwelling and work space, sameness of circulation and leisure, for an overall solidarity between humans that would erase class boundaries. The aesthetics of mass equalization, whether in cinema or in buildings, would, according to modernist prophecy, replace elite cult by a new aesthetic that would be truly revolutionary. So Benjamin believed with Le Corbusier.

Where there is the loss of enchantment there is the desire for reenchantment and Benjamin did understand that mechanisms of mass reproducibility might have the liability of generating mass seduction, charisma, cult of personality.[4] Such films would seduce, adulating the charismatic, fascist personality through seductive camera angles and larger-than-life settings. The student of film may recall *The Triumph of the Will*, which opens with the descent of the führer's plane from Wagnerian cloud to Nuremberg middle earth for the 1933 Nazi rally there. It is now known that Leni Reifenstal staged that rally—designed and orchestrated its graphic format—so that it might be filmed by her and then broadcast to the eyes of millions of willingly seduced Nazi sympathizers. When certain parts of the rally did not come out well in the rushes, scenes were reshot on a sound stage. Riefenstal was taking no chances; everything would have to march with the precision of the clock and power of the jackboot, soar in the tenebrous swirl of her camera, hyperconfidently claiming historical ascendancy and necessity. The führer—shot from elongated, dynamic angles merging him with the crowd below and the sky above—became in that film a heroic, mythological figure, a force originating from everywhere and nowhere, coincident with spectacular reality. This composed world could only appear on film. Hysterical commitment is its effect. Film thus, in its own way, may create the star, who is entirely a retrograde figure, carrying the aura of the aesthetic cults of the nineteenth century into the charisma of the present.

As David Ferris says in a book on Benjamin:

> There is the very real possibility that aura will be reproduced in and by the very media responsible for its "decline." What is clear from Benjamin's discussion, even though he does not say it in so many words, and what has become increasingly evident ever since, is that aura thrives in its decline, and that the reproductive media are particularly conducive to this thriving. . . . The star and the dictator had a similar function and origin. In both, the "amorphous mass" could find a face and a voice that it might call its own, or if not its own, that it could at least recognize and use to secure its own position.[5]

Since film aura is for Benjamin a matter of public seduction, film's revolutionary potential consists in its refusal of this charisma, its montage effect. There is no discussion of Sergei Eisenstein in Benjamin's famous essay "The Work of Art in an Age of Mechanical Reproducibility" but that would have been the sort of thing Benjamin would have had in mind. Film that refuses the politics of mass enchantment, that eschews the cult practice of star formation, that says "no" to aesthetic seduction, that turns its nose up at the sweet perfume of nostalgia, film that instead directs the human imagination to a vision of a possible future through formal experimentation in the telling of story: this is revolutionary film. Benjamin had visited Moscow (which proved disappointing for unrelated reasons), he was closely aware of surrealism and had interviewed Brecht (who refuses the aura of the actor by having the actor confront the audience). These were the forward-looking things.

Diana might seem tailor-made for Benjamin's retrograde idea of aura in an age of mass reproducibility. She carried the aura of the film star, and there was certainly a cult formed around her. Diana was a royal/upper-class/elite beauty cast into the world of the media in the guise of religion. Diana (the persona) was no doubt the result of the commonwealth world's longing for enchantment, upper-class charisma, a peep into the forbidden city of Buckingham, and a desire to wed glamour, celebrity, tabloid with film and royalty in a unified cultural delusion. People's princess yes, but in the form of royal slumming and royal identification that began and ended with media connection. Retrograde indeed!

I think this negative reading of the Diana cult is correct but also one sided. And not only because it leaves out the human factor, her pain, her place in the world as a woman in search of grace but also because, as a reading of the aesthetic sources of (her) cult, Benjamin is mistaken. Diana's aura does not derive entirely from studio buildup or royal hairdressing, a bevy of publicists and biographers. It also comes from cinema itself and the star qualities cinema creates as a medium. How do we understand, if not through the concept of aura, the halo exuded by the star in films? Benjamin is too quick in his dispensing of the film aura. And yet he also seems right. In the case of a medium like film where there are no originals, only prints/copies, how could one presume to stand in front of it, to be present to it so as to invest it with an aura? What is the "it" we would stand before if not the screen, which holds nothing but light on its surface? How could we think ourselves present to the figures and scenes projected on screen? What kind of "being present" is this?

Here aesthetics gives way to the conundrums of philosophy. A good place to begin in addressing this most vexing conundrum is with an early essay on cinema by the art historian Erwin Panofsky, "Style and Medium in the Moving Pictures"[6] This essay, which is full of important insights into the medium, concludes with a surprising assertion:

> The medium of the movies is physical reality as such: the physical reality of eighteenth-century Versailles—no matter whether it be the original or a Hollywood facsimile indistinguishable there from for all aesthetic intents and purposes—or a suburban home in Westchester; the physical reality of the Rue de Lappe in Paris or the Gobi Desert, the physical reality of engines and animals, of Edward G. Robinson and Jimmy Cagney. All these objects and persons must be organized into a work of art. They can be arranged in all sorts of ways ("arrangement" comprising, of course, such things as make-up, lighting and camera work. . . . The problem is to manipulate and shoot unstylized reality in such a way that the result has style. This is a proposition no less legitimate and no less difficult than any proposition in the older arts.[7]

The medium of film is physical reality as such, meaning the desert stones of Monument Valley, the harsh sun, and the facial grimaces of the smoldering John Wayne (Ethan Edwards), the materials of the past that compose it. Of course the film medium is from one point of view nothing but light projected on a two-dimensional surface from film stock (or analog tape or digital formats). But Panofsky does not mean this by his use of the word *medium*. His use of that word is imported from the study of the visual arts to refer to the stuff the thing is made of, which shines through in its realization. When it is said that the medium of Michelangelo's *David* is marble, this means that marble is the stuff from which Michelangelo composes, the stuff whose "form" he liberates from within the stone. Marble is there in the finished product in every possible way. The kind of sculpture that is made is conditioned by it, its expression inflected by the materiality that shines through. The gleaming surfaces of Bernini, the smooth, lyrical fleshiness of Canova sculpted from *pietre dure*, hard stone, are tours de force in marble given what that material is. Change the material and everything changes with it. The materiality of the finished form is something that cannot be abstracted from visual experience or from meaning and effect. That Bernini can render the hard clarity of marble soft, making hair bounce, is a work of magic. Were he a modernist working in string and wood, it would not be so hard, nor make us catch our breaths. These things give truth to Hegel's adage that "not all things are possible in all media of art" and, related, that it is the discovery of the potentialities of any given medium, their exploitation and, indeed, creation that defines the history of an art form.

Cinema was from the beginning, Panofsky tells us, a lively game of discovering what worked. It quickly gravitated to things that move (train robberies, comic incidents with gags), close-ups (the expressivity of the face), cutting across locations (transporting the viewer aesthetically from one location to the next in the way theater could never do), moving between past and present (the flashback), and exploiting human and natural physiognomy for expressive values in a way that called forth human temporality (desire—the future, remembrance and being marked—the past). These discoveries began to take shape within a few years of the introduction of the Edison box, which made film possible. Panofsky should know; he was there in the back rooms of Berlin warehouses in 1905 watching two reelers. He writes from experience, having been there "from the beginning."

When Panofsky asserts that the medium of the moving pictures is physical reality as such he is speaking to the central role of physiognomy. It is Garbo's eyes, Jimmy Stewart's twitch, Gary Cooper's tight lips, John Wayne's swagger that are the substance of their cinematic expressivity. Catherine Deneuve's icy classical visage is pure marble. Even the voice becomes what Cavell, following Panofsky, calls part of the "individuality" of each of these stars. We listen for the peculiar twang of Stewart, more high pitched, breaking slightly in the upper registers when he is anxious; for the low-throated ripeness of Joan Greenwood's high-Ealing-studio Englishness.

Panofsky was a creature of silent film, where there is no screenplay, only story framed by snippets of plot and dialogue that appear from time to time on screen. In silent film story is a frame that is realized essentially through gesture, close-up, editing, camera. The actor's exaggeration of mouth, her slow gyration toward the center of a room, her turning of the eyes spoke volumes on screen—spoke all that cinema could speak in that medium. The star drew out her gestures, prolonged them to make them sing like an aria. Everything happened visually, making silent film almost a different medium from sound film. Panofsky, in wishing to define film through contrast with the theater, found an ally in silent film where the screenplay is nonexistent.

Panofsky clearly believes screenplay secondary to what happens on screen visually. In spite of Panofsky's failure to properly value the screenplay,[8] his principle that in film the actor's *physiognomy* is central in creating the character remains a good one. A screenplay adds the critical dimension of talk, and this talk must be integrated into the visual flow of the film. Talk demands its own physiognomic realization—it has rhythm, inflection, idiosyncrasy that must synergize with visual physiognomy into narrative. A good screenplay is written or adapted with this in mind, and a director is in the business of syncing visual physiognomy with the rhythm of sound. This is what prepares the distinctive feature of sound film comedy for example: screwball films where Cary Grant yaps at Rosalind Russell, who yaps back in a way that sets the pulse to their superanimated faces. How an actor talks is part of how they look, the overall feel of their presence. Sound contributes to their screen presence, which returns us to that most striking of Panofsky's remarks: the medium of film is reality as such, which makes it seem that an actor's presence on screen is, for this art historian turned film critic, that of marble in sculpture: right there to be touched, right there to be heard. The actor is to cinema as marble is to sculpture.

Except that this cannot be right. For one thing, marble is inert while the actor dynamically participates in the making of his or her character. We have famous stories of Hitchcock treating actors like things, positioning them as if blocks of wood or statues. His film *Vertigo* (1958) is about a man fixated on such manipulation of women, which takes the form of necrophilia. The protagonist aims to recreate the dead through the living from shoe to hairstyle, then to possess her. The film is a metaphor for cinema generally. Actors in fact have a strange status between becoming thing-like and remaining persons on screen. They act, are *active*. And yet they exist in a film as appearances, bugs under glass, objects of porcelain. The star is fixed and free at the same time, figure and figurine.

But the real conundrum is that marble is physically present to the viewer in a way the actor definitely is not: you can reach out and touch the marble, just as you could jump on stage and touch the theater actor. But you cannot reach into a movie and make contact with Grant or Russell. They are not in this immediate sense present to you. Then what kind of material presence is this that causes Panofsky to use the metaphor of an actor shining through the screen like Michelangelo's marble through the sculptural form?

It is worth reciting a remark of Stanley Cavell's (from *The World Viewed*): "The reality in a photograph is present to me while I am not present to it; and a world I know and see, but to which I am nevertheless not present (through no fault of my subjectivity), is a world past."[9] One might take Cavell's remark (a riff on Panofsky) to mean something like this: central to film is that although not literally present, the actor (or mountain or field and stream) *seems* present to us. It is only that we are absent from him (it). Cary Grant, is, unfortunately dead, even as I watch *The Philadelphia Story* (1940, directed by George Cukor) I am certainly not present to him—not able to address him. And yet there is a strange way in which he *seems* present, strange in a way not true of his oil portrait, nor of any sketch of him. "Look," I exclaim with delight, examining a photo from an old album dug up from the attic. "It's Uncle Harry, how young he was then, the ruddy cheeks, swift glance of his darting eyes, hair parted in the middle, the dashing figure. It's as if he were with us today." But now is many years later; I am revisiting a tattered home movie or perhaps a black and white photo in a brass frame. Harry is long departed, yet I've the sense that time has collapsed and he is with me, or I with him. The photo seems a window into the past or a way of bringing the past to the present. Roland Barthes says photography is a memorial to the dead, a way of making the dead present again among us while also confirming their absence, since the photo reminds us it was *then* that we knew them (and no longer). This ontological disturbance, this sense that the barriers of time are being occluded, associates photos and films to dreams, séances, to the intensity of memories unfolded in mesmeric absorption, as if memory were a film replayed within the self. As a child I had an overwhelming belief that I could replay my life at will, and I would sit in bed reviewing and rewinding the course of my life in memory's chamber, each time speeding things up or slowing them down, depending, and all with the most intensely visual sense of their presence. Time gained was also time lost, since the automatic replay of my own past confirmed its pastness. I was passive before my own memories, unable to enter them, able only to allow them to consume me.

Cavell calls the world in a photo or film a displaced form of presence: Garbo is present to me, but I cannot be present to her, I cannot speak to her in the film, nor reach out and touch her. Then is Garbo, who was once physically present to the camera (at the moment she was filmed), really there before us, as if directly out of the past? What kind of *there* is this?

The question of the film aura resides in the answer. One must find a way of understanding how things are present to the viewer on screen without falling into the trap of saying something absurd or unintelligible. For the aura depends on something present (marble, the Villa D'Este, the burning bush). How then shall the presence of things on film best be described if not like that of sculpture?

Certain philosophers, most notably Kendall Walton, have argued that physical reality is *literally* there to be seen in a photo or film. According to Walton film is neither a copy of the world (in the sense that a drawing of the Vatican is a copy), nor is it a representation of the world in the sense of a painting. Photos are transparent windows to that which is no longer there. In his words, "We see the world through them."[10] A historian comes across a photo of the Lower East Side in the archives of the New York Public Library, finding herself dazzled by the teeming spaces of tenement, fishmonger, rabbi, and lorry. A window into the past indeed, she thinks. What luck to have secured it. Photos are, Walton tells us, mirrors reflecting back into the past—strange mirrors, because what we see in them is no longer there. That the thing that caused its "reflection" in the photo is a thing of the past makes our seeing of it, in Walton's phrase, "indirect." "To look at a photograph is actually to see, indirectly but genuinely, whatever it is a photograph of."[11]

This is because photos are counterfactually dependent on the things photographed, something not true, Walton says, of paintings. A painting may represent a person whether or not that person was the actual source of the painting (sitting in the studio and being painted). A painting of Winston Churchill may be a perfectly legitimate painting of Churchill, whether or not it was painted in the presence of Churchill. The painter may never have seen Churchill at all, except in photos and newsreels. But a photograph cannot be a photograph of him unless he was the man caught in the lens. A photo of another individual dressed up to look like Churchill is not, properly, a photo of Churchill, but, instead, a photo of whomever was snapped that *pictures* Churchill. This is the difference between photos and pictures. Photos are dependent on the reality they represent, as vision is. In order to say that I see an elm tree, the tree I am regarding has to actually be an elm, not an oak or maple. Similarly, in order to say Cary Grant is in the mirror, it has to actually be him reflected, not a look-alike. This is how vision, mirrors, and photos differ from paintings and form a category of thing. Another way to put this is that the relation between photo and world is in essence mechanical. Photographers stylize, yes, they frame their subjects by positioning the camera, choose film stock, set the time of exposure, the light meter, and so on. These choices generate perspective, mood: art. But, once these choices have been made, there is no more human intervention. The camera goes snap and that is that. This is not true of oil painting, where the entire act of painting is human choice, a human intervention. (Let us avoid the dilemma of Photoshop, which convolutes the very medium of photography by turning a photo into a baseball field of cut-and-pasted images from the computer file and the World Wide Web).

Walton's aim is to identify this larger category of perceptual experiences that share the property of counterfactual dependence. The category includes mirrors, telescopes, microscopes, and, he thinks, photos. All these devices, he believes, share the property of transparency. We see through them to reality. Walton seeks to demonstrate that photography is a case of transparent perception by assembling examples of seeing through space meant to prepare the claim that photos allow us to actually see through time to the past. The simplest example of visual perception is two people each looking directly at the other (in a café, across a crowded room). One can then move on to such examples as observing a suspect through a one-way mirror, a ship through a periscope, the human heart through an ultrasound machine, a cellular microcomponent through an electron microscope, a distant star through the Hubbell telescope. In each case the gap widens between perceiver and perceived, becoming less and less reciprocal, more and more "distant."

Walton's leap is to move from these cases to seeing across time, which is best imagined, Walton tells us, through the experiment of looking through a "multiply mirroring device." The device (on this scenario) reflects the world through so many mirrors—each mirroring the next—that we lose all orientation as to where the object perceived is with respect to ourselves. We can't trace the figure seen through this hall of mirrors back to any particular location. The world seen through this multiply mirroring device is presented to perception while also screened from it. The example is meant to warm us up to the idea that seeing across time is but one stage further than seeing across space (through the device). Seeing across space can also eclipse perspective, derange orientation as to source.

It is important to note that behind the multiply mirroring device is still the sense that the thing seen in it really does exist somewhere out there, right now, in its material robustness: it's just that we can't pinpoint the object's whereabouts through the device. There is a route one could take to the object perceived; it is only the device that fails to yield it and in fact obscures it. This is quite different from saying we can see through the photo or film back into time, where there is no possible road map back to the thing that we could take to arrive at the past. In looking through a photo toward the past, our estrangement from that past is metaphysically permanent.

In his 1984 essay "Transparent Pictures," Walton finds no particular problem with this: "I don't mind allowing that we see photographed objects only *indirectly*, though one could maintain that perception is equally indirect in many other cases as well [such as mirrors, images produced by lens, etc.]."[12] For him, photographic transparency is no more problematic than the transparency of a mirror. This is unpersuasive. Seeing into the past is a concept a hundred times more bizarre than seeing through a mirror, which is not bizarre at all, or even seeing a fetus through an ultrasound machine.

The best case for seeing into the past is Walton's example of seeing a degenerate star. One looks through the Hubbell telescope to a star, but by the time the light from that star reaches the telescope the star has in fact disappeared. It is in the past. We are seeing, in this sense, that which is past. But note: here we are seeing it directly, through its original light source. It is just that this light has taken a long time—too long a time—to reach us. The case is different from looking into a photo. The star appears to us in its original light: the light that traveled from it to the device, just as a person appears reflected in a mirror through the original light that ema-nates from the person. This is what makes the person "present" in the mirror. Seeing something or someone in a photo is not seeing them in the original light that traveled from them to the photo-graphic device. That happened long ago and led to an imprinting of the image (like a sound recording) onto the photographic emul-sion.[13] At a later time the film emulsion has been developed and at a still later time that film has been projected. We see the original imprint through new light. That is what we see; the imprint and nothing else. A photo is, Cavell says, a mold taken of reality, that is, not a mirror of it. One does not see through the photo to some-thing beyond (where? what? how would you get there?); one sees into the photo.

If one wants to call this indirect seeing, well and good, but one has then changed the language, for this is emphatically not the kind of "indirect seeing" one has looking into a mirror. We see into the photo, and then, knowing that it is a mold of the past, have access to the past without quite seeing it. One feels the presence of the past in a photo because it hovers, elusive and unseen, in the man-ner of trace or aura (that which signifies that which is not). In this felt conviction resides the strangeness of a photo, in its aura.

That we look *into* photos and films rather than through them is the backdrop for Jean-Luc Godard's film *Band of Outsiders*. In the middle of that film, in the middle of that war, a soldier stumbles into a cinema, fully armed and dangerous. He sees a blonde bombshell on screen and, mesmerized, draws closer and closer to the screen, which looms larger and larger as he approaches it. The blonde's bombshells are set to go off right in front of him; her curves are get-ting slinkier as he approaches. Finally, overcome with excitement, the soldier tries to hurl himself into the film screen where paradise, he is sure, may be found. Naturally he fails. And yet his fantasy is ours: that what resides in the world of the film has substantiality when it is nothing but imprint, apparition. Film is the stuff dreams are made of, and it makes us crazy, like moths to the flame. Here is the source for the star icon: that we almost see the reality that hovers beyond, almost see into a film as if into a mirror, but do not, it being an illusion. Here is the source of our fantasy: seeing should be oceanic in its capacity to reach things that are mere apparitions while carrying the perceived force of substantiality. It can drive a soldier, already crazed from battle, wild.

I should like to call film an intermediate case, about which it is too simple to say either we see the past in it or not. In saying this I am recruiting a thought of Wittgenstein's, the relevant passage being from his *Philosophical Investigations*:

> A main source of our failure to understand is that we do not *command a clear view* of the use of our words.—Our grammar is lacking in this sort of perspicuity. A perspicuous representation produces just that understanding which consists in "seeing connections." Hence the importance of finding and inventing intermediate cases.[14]

It was an important part of Wittgenstein's teaching at Cambridge University in the 1930s and 1940s to try to "find and invent" examples of objects about which it was too simple either to say it was a "p" or a "not p." He was interested to explore items about which it was, or seemed, too simple to say it is or is not a number, is or is not a person, is or is not alive, is or is not a game, is or is not language. Film perception is exactly this. It is too simple either to say we see Grant on screen or we don't. His presence is absolutely peculiar. (Walton is almost right.)

A medium of art sustains absorption, makes magic through its peculiarities. The film medium is peculiar in its mode of making the world present to the viewer. Whereas sculpture sets materials directly before you, film does not. And yet physiognomy is "there." Film is counterfactually tied to its sources, and yet we cannot see them (but we almost seem to). The things happening in film seem utterly present, more real than real, and yet also feel of the past, spectral. Caught between presentness and pastness, film is neither quite. This double experience of things present and things spectral, of things that seem completely present and also strangely unavailable, convulses the imagination, which is unable to sort out its relationship to things screened, to their mode of being-there before us. In this resides film's aura.

Without cinema there could be no star icon as we know it. For her position before us is cinematic: she is present, more real than real, and also effervescent, of another galaxy. Film creates the star through its aesthetic position. Through film the star is born.

six

Stargazing and Spying

ALTHOUGH SOME of the greatest film directors have avoided stars (Tarkovsky, Bresson, Malick), film usually converges around the star, and I mean *Casablanca* fashion, Bogart and Bergman fashion, Claude Rains and Sydney Greenstreet fashion. The star is an object of glamour that the studios invested millions in building up: about this Benjamin is right. But the star's magic does not reside entirely in external marketing. Star aura originates elsewhere, in the conditions of film perception that transpose her presence into that of a being suspended in a distant galaxy, magnified through the film frame, dangerously present, hauntingly absent, burning brightly from a vast, cool distance. That is what allows the studio to accrete her and controls how she is marketed. Star image is the marriage of cinema values and distribution values. The star system runs on this. And the life of the icon runs on it as well: film aura, limited public access, and controlled removal from the public. Only when Garbo's career was over was she at liberty to wander the streets of midtown Manhattan, dragging her dull overcoat and carrying overstuffed packages. While under the sway of the studio, while casting spells on the public, she was kept apart, and under, as it were, lock and key (call it a studio contract).

Garbo wandering the streets of Manhattan during daylight, shopping for curtains or a new gray skirt, is one thing. But on screen the room is dark. Her bed is bathed in phosphorescent, silvery light, the pillow white. In this study of contrasts she stands out as silver and satin, but indistinctly. Her assistant walks toward the bed.

Then the camera approaches her and her long, lithe form becomes visible, stretched across it. She faces us, imploring, mournful.

"Has madame slept well?" the assistant inquires.

"No, I've been awake thinking."

"It is time for the performance." The assistant speaks in a Russian accent. When she is done speaking she turns on a light by the bedside. Suddenly madam is bathed in whiteness. We first see her in profile, then in that baptism of the star, the close-up.

"I think," Madame wails, "I've never been so tired in my life." Her face then fills the screen, eyes pools of dampened light, cheekbones taut against the angular beauty of her face.

This is *Grand Hotel* (1932, directed by Edmund Goulding, screenplay adapted by William A. Drake and Béla Balázs from the play by Vicki Baum), and she is prima ballerina facing the preternatural waning of the body that only a dancer may know, since it happens before the age of thirty-five. Mournful, aestheticized, petulant, overwrought, she is the very image of dolorous perfection, of limpid glissando.

Two and a half decades later the same lights will be turned onscreen in triplicate. This time the scene opens with camera panning around the closed courtyard of a mundane New York apartment complex, peeping from mid range into the glass box of each tiny apartment to see what is happening within. Close but not too close. This is, we believe, L.B. Jeffries's point of view, since he is the one with the broken leg who stares incessantly out the window, studying his fellow apartment dwellers like bugs under glass. But when the camera completes its circular tour and turns into Jeffries's own apartment, we see him asleep; the point of view of the camera panning around these apartments could not possibly have been his. He is being observed close-up by another, which is shocking, as if this invisible person has stolen his point of view, claimed his consciousness. This will be the battleground of the film, fought over the topic of point of view, authority of gaze, lifestyle, control, in short, marriage. We do not know it yet. We wait a second in anticipation of whose this invisible source of consciousness is. We wait only for a second when her shadow appears, preceding her, gliding over him in a style mysterious, sexy, just possibly sinister. Only then do we see her classical blonde head. It bends over him until the camera angle impossibly switches and she is bending out of the screen toward *us*. Larger than life, her eyes glow like halogen lamps (or camera bulbs) with excitement; they careen toward us until we feel we shall be devoured by them. It is effervescently scary, this uncanny emergence from the screen toward us that should not be happening but is, unnerving until the camera performs a switchback and shows us her lips touching his, shot from the side. It is *their* kiss now, not ours. We are relieved and robbed, overprepared and let down—and this for the second time in Hitchcock's tour de force of filmmaking, since the first thing stolen from us in this less than two minutes of filmmaking was our belief that the vantage point of the camera was his. He in turn has been robbed of point of view and kissed, a declaration of intent (hers) to steal something from him, render him passive, feminine, immobilized. The kiss is rapturous, their heads utterly filling the screen, their lips joining and gently pulling apart. One of the best in cinema, it is also part of a dialogue between them, she stealing up on him and doing things, he in a secret tense enjoying the passivity.

Then they talk:

"How's your leg?" she asks.

"It hurts a little."

"And your stomach?"

"Empty as a football."

"And your love life?"

"Um . . . not too active."

"Anything else bothering you?"

"Mmmm. . . . Who are you?"

Her voice is sinewy, balmy, gentle in the way it caresses his body. The voice lingers as she continues:

"Reading from top to bottom, Lisa, Carol, Freemont." She speaks each of these words as she slides through the apartment turning on lights; one light for each word. Her white chiffon dress swirls about in its own ballet, liquid, ecstatic, in complete control. She has stated her name, but as the endpoint of an action that is also

FIGURE 6.1. Jimmy Stewart, Grace Kelly, and Alfred Hitchcock. Bettmann Archive, Corbis

her source: a self-illumination through light and movement. This is the bright-lighted presence of the star, the perfect harmony of appearance and word. She is dulcet flesh and filigree of light.

This debate about who she really is will remain unresolved to the end of the film. There Jeff sits in the final scene, immobilized (two legs busted now), his fiancé (Lisa Freemont) sprawled on the bed next to his chair, reading a magazine meant to prepare her for the rigors of his life in the rough (tough macho journalist) traipsing through Burma and Tibet. In fact she is reading a glamour magazine tucked secretly inside the other: she has her own ways of exercising autonomy—and also control over him. His claim to control her has failed, which is what he hates about her and also what makes him crazy with desire for her.

Jeff is a photographer who stuck his neck out too far. He's in a wheelchair because he rushed onto the track of an auto race in order to get the perfect *Life* magazine picture. In the course of staring out the window of his apartment at the other apartments around him, he will claim to have uncovered a crime—ironically he will be correct, although the interpretation was made without evidence and as a form of self-projection (of his own murderous instincts) onto life across the way. Jeff's delusion is that he can solve this murder without being discovered, that he can remain in the safety of his apartment watching, the perfect voyeur. Lars Thorwald, the murderer, played in the hulking form of Raymond Burr, eventually figures out who this Peeping Tom is and comes after him. And how does Thorwald figure this out? Because Lisa, teeming up with "Jeff" to solve the crime, decides to break into Thorwald's apartment to recover hard evidence of the crime, which turns out to be the wife's wedding ring. Thorwald returns while she's in his apartment, challenges her, possibly threatens (we cannot tell, nor can Jeff, since it is happening across the apartment complex and the sound does not carry). Lisa flashes the wedding ring at Jeff, showing it is on her finger. Thorwald sees her do this and looks across to see where she is pointing her finger.

When Thorwald enters Jeff's apartment, soon after he demands, "What do you want from me. Is it money you want? I haven't got any. Speak, say something!" Jeff says nothing. It is the murderer who is sympathetic, not he. Jeff stalls this hulking frame as it approaches him, threateningly, by shooting of flashes at Thorwald, each momentarily blinding him, stopping him in his tracks. Thorwald asks, "Can you get the ring?" Jeff answers, "No." Meaning it's too late. Thorwald throws him out the second story window of his apartment into the courtyard he has been studying under the illusion of impunity.

If there is one moral of Hitchcock's *Rear Window* (screenplay written by John Michael Hayes adapted from Cornell Woolrich's story *It Had to be Murder*), it is that photography is a shield for the deployment of fantasies of omnipotent desire onto the star from the presumed safety of viewing. The viewer of the star, as Jeff of Lisa, casts the star in a haze of unapproachability, while also projecting onto her all manner of possessive fantasy. It is this doubleness—wanting her and wanting her to remain unavailable because one is terrified—that produces the desire for star aura. Up close and personal is unsafe; possession should take place in fantasy only. This is the very definition of voyeurism, this combination of possession and reveling in unapproachability. It is Jeff's "photographic" fixation on Lisa as well as ours. The star exposes Jeff (and us) to danger if she is as lively as Lisa, which is to say as resolute at getting what she wants. That she's broken into a man's apartment to steal a wedding ring is hardly fortuitous: that she flashes it at Jeff in a flush of pride suggests she believes she's conquered him, married him by stealth. It is Jeff's wedding she's indirectly planned by consorting with him in this mad adventure. Talk about craft—she's the craftiest since Eve herself. When Thorwald asks Jeff if he can get the ring back and Jeff answers no, this is a description of Jeff's own destiny. It is too late: he's hitched. Will Jeff too become a murderer?

We too are hitched to this Hitchcock thriller since we share the voyeuristic position with Jeff, something with which Hitchcock hits us over the head by unhinging her from the screen and having her gesture dangerously toward us in the early scene of the kiss. By orchestrating the aura of the star to contain hidden threats (for Jeff and ourselves), Hitchcock breaks through the safety net of the voyeur, thus showing how deep it goes in the experience of film, how central it is to the very making of the film aura. The film aura is not only a patina of the past, it is a living projection of voyeurism onto things and, here, stars, who are shielded from us through their distance, their pastness. When they break role into the present, they terrify. This is Hitchcock the philosopher of film showing us what a film is, what the aesthetics of viewing are all about, what star aura consists in. Film licenses the omnipotent fantasy of possession (of having Grace Kelly) because, and only because, this cannot take place, given her status as film object, denizen of silver screen rather than middle earth. Lisa Freemont (Grace Kelly) is the classic Hitchcock blonde: at home in her aura and not at home in it, there and not there, wanting to be an idol and wanting to be real (to him, to us). This is the unstable character of the film aura: that it arises through fantasies of control, through the desire to possess the star and also revel in her inaccessibility, and through the safety net of the theater that makes things appearing in two dimensions liable to this form of projection. A great deal has been written about Hitchcock's women. Whether a beast caught in a cage (Marnie), a blonde in a motel shower (Marion), a woman regarded as rotten fruit by a fruit seller who uses a necktie like a noose (*Frenzy*), or a frozen social dilettante whose game of bird giving excites a man's birdlike mother to metaphysical derangement, simultaneous with the sky falling in (*The Birds*), the dangers they encounter are of a piece with those they generate—and generate through unstable patterns of distance and closeness, inaccessibility and desire. By systematically breaking through the shield of the aura to inquire into its sources, Hitchcock shows us why, on a Saturday afternoon when the sun is shining and birds are chirping in the trees, we should choose to enter a dark theater, sit immobilized (as if in a cast), and watch some blonde get stabbed in the shower until her blood flows down the bathroom drain and her dead eye stares back at you like a horrifying mirror of your own murderous fate. These stars bring danger, which is inevitably associated with their being seen by men, with their being viewed. The key to the Hitchcock star is this (black) magic of appearance.

The star is a bird in a cage but also something let loose, which is how the viewer, like the male protagonist in the Hitchcock film, appreciates her and what he also cannot stand about her. When Garbo is lit, she is lit by others. Her aura is one of dependency. Lisa Freemont lights herself; her art is formidable. She is a conjurer aware of her powers of manipulation. Grace Kelly will act in five films before giving up the enterprise and turning to light the Kingdom of Monaco. Diana is given an aura without acting in any film. She wears it in the public imagination, a crown she did not solicit that she does not understand. This is not true of Grace Kelly, who remains spunky and unexpected, who does what she likes (chooses the role of queen, etc.). In an excellent essay on *Rear Window*, film theorist Tania Modleski says:

The film gives her [Lisa Freemont] the last look. This is, after all, the conclusion of a movie that all critics agree is about the power the man attempts to wield through exercising the gaze. We are left with the suspicion . . . that while men sleep and dream their dreams of omnipotence over a safely reduced world, women are not where they appear to be, locked into male "views" of them, imprisoned in the master's dollhouse.[1]

Hitchcock has fixed Lisa as a bug under glass, then demonstrated to viewer, photographer, and director alike their failure to pin her down. This is the star, her aura is a cage, but also a sign of her placelessness (we do not literally see her because she is not finally anywhere). Her halo is a projection objectifying her as well as a signal of her inability to be owned and occupied by the viewer, who believes himself her omnipotent master. That viewer, like Jeff, is the passive one, the one immobilized. It is through this license to fantasy combined with an ongoing sense of inability that viewer absorption is sustained. (Many a marriage will sustain itself on these terms.)

These are the terms of audience attraction, contradictory, fantastical, desiring, permanently stunned. It is how I feel every time I watch Grace Kelly in this film, and I could watch her a thousand times. This structure of attraction in cinema is what prepares the existence of the star icon outside the silver screen. She is unapproachable, distant, a bug under glass and yet a bird trying to get out. The structure of our voyeurism, fascination, identification, omnipotent fantasy of desire and possession, and, finally, powerlessness is critical to cult formation around her. The life of the icon is as it is because her public are in actual solidarity with her, but also because they are voyeurs of her. She must seem ours but never come too close. That is dangerous, destructive of her halo, of our desire. Ours is a desire that never wants fulfillment, but wishes to remain always on the verge of it and always within the fantasy of having already achieved it. The structure of the viewer's position in watching a Hitchcock film is importantly transposed onto the way a public views their star icon: Diana, Jackie, Marilyn, Grace. This is why, as I said earlier, the drama of power is also one of mutual powerlessness. It is also why the star icon must carry the aura of a film star: so that we can exist in this state of cinematic voyeurism in relation to her. The star icon is who she is because of things beyond her control: her physiognomic presence, her life story, the way the media has gravitated to her, the way the public will not let her go. But the public wills itself into Jeff's position and the film viewer Jeff resembles (which it is Hitchcock's genius to point out). This is a position in which the omnipotent fantasy of possession and identification is really a disguise for willed powerlessness before the star icon.

Diana was a Garbo: liquid, imploring yet also self-contained eyes, passive, a figure of radiant dependency from *Grand Hotel*. But she was also a Lisa Freemont, a Grace Kelly refusing to stay put. The public reveled in her derangement of the royals, who were by the 1980s all too shielded from public perception and not well liked (Charles being the case in point). That Diana proved a volcano in the house of Charles, refusing the role of bug under glass, was every bit as fascinating to the public as Lisa's derangement of Jeff, who like Prince Charles is a less than likeable character. And Grace Kelly? She played herself, classical, seductive, acting as she pleased, ready to move in with her Mark Cross handbag to the house of Monaco, gambling with life and, until the end, winning.

The stance of film voyeur before star-in-movie is in important respects the same stance as that of awestruck public before the star icon. It is a stance of safety, which, if broken through, produces danger and may break apart. It is a stance of control while powerless, controlling while powerless before the Lisa Freemont who breaks out; it is a stance licensing fantasy and identification while also retaining the absolute barrier of distance. What film theory says about Lisa is in this respect true of Diana and Grace, the star icons offscreen (in the world). Before Princess Diana, the public played kitchen maid looking through the keyhole, peon before queen, while she played Garbo and Grace Kelly in the house of Windsor. I think Marilyn and Grace, being actual film stars, were viewed more erotically than Diana and possibly Jackie, both of whom were held in awe, placed in temples of religion. All were, however, objects of a voyeuristic stance.

This is how the structure of film viewing prepares the cult relation to the star icon, to Diana, Marilyn, Jackie: by developing a gaze
of identification, admiration, distance, control, and powerlessness.
Or at least this is how film prepares *half* the public's gaze on its star
icons. The other half comes from tabloid and television and is the
gaze intimate with her soap opera life, of a fan rooting for a good
ending to it, of a Roman public wanting even more blood and guts.
We shall turn to this soon enough.

The position of kitchen maid before the keyhole, of voyeur before queen and vassal before princess is what one of the greatest
films ever made is all about. That particular film's story of privilege
versus prying, of privacy versus public, and of "what the kitchen
maid saw through the keyhole," to quote one of its lines, was made
half a century before Diana became the people's princess. And yet
the film is as prophetic of the star icon as it is revelatory of film per
se (and the film viewer). I speak of *The Philadelphia Story*, adapted
from the play by Philip Barrie written in the 1930s to orchestrate
Katharine Hepburn's comeback from "cinema box office poison"
to star (adapted screenplay by Donald Ogden Stewart). The film
pitches lordship against something finally stronger: a thing called
Spy Magazine.

"There's nothing in this fine pretty world like the sight of the
privileged class enjoying its privileges," Macaulay Connor says with
cynical ease in *The Philadelphia Story*. And who is Macaulay Connor ("Mike to his friends," "of whom," Tracy Lord says, mocking
him, "you must have many")? A writer without wherewithal who
is therefore slumming it, working for *Spy Magazine* to put food
on the table and a roof over his head, and who resents every inch
of Tracy and her upper crust life, *Spy* has sent him with his fian-
cée Liz, a photographer, on "undercover" assignment to get the
goods on the wedding of Tracy Lord (aka Katharine Hepburn).
The Lords are one of the oldest families in Philadelphia, "stink-
ing rich," and guarded of their privacy. He has entered their dazzling private world, and at the moment we find him dancing with
Tracy at the edge of her pool, also busy killing the champagne bot-
tle with her, enjoying a mad flirtation at three in the morning. It

is, not coincidentally, three in the morning of the June day she is
meant to marry George Kittredge, man of the people, political tim-
ber, up from poor mining to general management, in a wedding of
"national importance." Conner sprawls scornfully onto a poolside
chair, his feet up, chin out, and pontificates scorn about the party
he has in every way been party to on this shimmering occasion.
His speech elicits from her: "You're a snob, Connor . . . the worst
kind there is, an intellectual snob." Then she adds: "You made up
your mind awfully young it seems to me," lecturing him, with
the remark, "You'll never be a first-class writer or a first-class
human being until you've learned to have some small regard for

FIGURE 6.2 Katharine Hepburn, *The Philadelphia Story*. Trailer, Wikipedia,
public domain

human frailty." Except she doesn't quite complete the final word, realizing she's mouthing something that was more or less just said to her by her ex, Mr. C. K. Dexter Haven, aka Dex, aka Cary Grant who has shown up at the Lord mansion after two years drying out from an excess of alcohol and then working in Argentina for *Spy Magazine*, which is the ostensible reason he is back at the Lord manse.

The film is a study of queens, princesses, American royalty (as close to it as America can get anyway). It is also a study of the star system that hounds such people as she. The film begins with Tracy Lord's breakup from C. K. Dexter Haven two years earlier, showing him storming out of the house after "socking" her. Tracy has unmanned him by breaking one of his golf clubs with a smirk on her face as he stormed out the door and had clearly unmanned him throughout their marriage with her "withering glances." He drank too much. Now, two years later, he's back, sober and slinking around, not only working for *Spy Magazine* but also engineering the secret entry of Connor and Liz into the house on a pretext. It will take the entirety of the film to demonstrate what Dex is actually doing there, namely, saving her from that marriage. Connor, ready to jump to negative conclusions about people he doesn't approve of (of which there are many, whole classes of them), assumes Haven is there to get his revenge, and get it by scooping the wedding for *Spy* magazine (a magazine Tracy rightly detests).

Spy is a bellwether for the Lord family. Uncle Willie is first seen reading *Spy*—I should say devouring it in the manner of young boys absorbed in the latest issue of *Playboy*. Uncle Willie is a "pincher," a bachelor who lives for the frisson of pinching a girl's bottom on the sly and for little else. Dinah, the younger daughter, is crazed to find out the gory details of the adult goings-on from which she knows she's constantly being excluded. With an instinct for the unseemly, she's a preternatural master of the subtle art of spreading rumor, generating innuendo, and fomenting all manner of brouhaha. (A generation later she would be editor of *Vanity Fair* or similar.) It is she who spies Connor in a dalliance with Tracy in the early morning hours of Tracy's wedding day. It is she who sparks off a comic round of misperception through her certainty that Connor must have taken full advantage of Tracy. Dinah is a perfect *Spy* reader. "I love it," she blurts out earlier that day at the stables, "It's got pictures of *everything*." A regular stalkarazzi of the upper crust is she.

"Anything in there about the wedding?" Kittredge asks greedily, after Tracy has tossed Uncle Willie's copy into the stable dust. Tracy gives him what Dex will later call "the withering glance of the goddess." After cringing, he retorts, "Well I just thought . . . with you being one of the oldest families in Philadelphia and me getting somewhat important myself . . . it's just luck, but . . ." This produces a diatribe from her about such magazines, to which he responds: "People like celebrity," adding, "Suppose I took it to go into politics?" meaning they would live in the public eye then, and what would she do? She responds that under no circumstances will such intrusions as those of *Spy* take place in *her* home. "Don't you mean our home?" he asks. "I mean very much our home," she replies, about half meaning it. She has not worked out what that home might actually be like, with him getting somewhat important, singularly dedicated to having more of it, greedy for political power and the public celebrity that goes with it. Only at the end of the film's twenty-four hour period, after he's broken things off, do his true colors emerge. For he's ready to "let bygones be bygones" when he finds out that a Mr. Sidney Kidd is "here," at the wedding, having shown up to personally photograph it for *Spy*. Kidd was not supposed to be there, but when it's known that he is, Kittredge's responds, beaming, "Well Sidney Kidd's presence makes this of national importance." National importance is, we now realize, what he craves, and he's willing to marry to get it, since he can't tell the difference between love and ambition. When she rejects his truce, he storms off, muttering, "You're on your way out, the whole lot of ya," meaning the whole upper crust as well as the class of persons who will not get into bed with *Spy*—even over matters of "national importance." Perhaps that is the matter of national importance, this question of who will bed down with whom for media purposes. Kittredge wants the whole deal: she on an embossed silver platter to admire, a trophy wife, a genuine "princess" and, moreover, someone who will give him the right profile for *Spy Magazine*. He's a stargazer, social climber, and political savant all at once. And he is just possibly correct about what the public wants of their lords and masters, namely, their presence in *Spy*, the ability to spy on them like the kitchen maid through the keyhole.

Tracy carries the aura of a queen, which is how she stuns her men. They're all gaga for her, the unapproachable, magnificent Tracy. To hear them talk is like hearing the Earl of Spencer sing praise poems to his older sister. They idealize her, *idolize* her. It's like marrying an icon you place on a pedestal where she may lord it over you in the silvery moonlight, statuesque in her frozen perfection. Not a recipe, Dex alone knows, for human contact, much less for a sustainable marriage. And diametrically opposed to the gaze of kitchen maid through the keyhole, of leering and scandalizing: the gaze of an Uncle Willie (a pincher) and Dinah (a child obsessed with its "pictures of everything," the more sordid the better). Dex is the only man to have gotten over the goddess thing because he hit the dregs on account of it, and took two years to pull himself up. Now he wants to get her over it, over this stargazed, dehumanized position: for reasons unknown, probably because he believes they could still make a good couple, an exemplary couple, so long as she takes the plunge as he has had to do. His humanization of her will take the entire work of the film, being about to fail at every moment, but, like their remarriage, is worth the risk. She will flirt with Mike, who will propose to her, she will drop Kittredge, audience assembled, wedding music playing, then finally Dex will come to the rescue, will find the moment propitious for a proposal, will risk humiliation or despair and pose the question. Then and only then (when she is about to glide down the aisle to remarry Dex) will her father say that she looks a queen. Then and only then will she rapturously reply, "And do you know how I feel? Like a human, like a human being." To which her father will then and only then reply, "And do you know how I feel? Proud." All is forgiven between them, in spite of his middle-aged dalliance with actress Tina Mara and her general disappointment with men. She has become, finally, a woman, meaning a human being—meaning, Lordly. Stanley Cavell, the master of all discussion on this topic, will call this the Shakespearean element, this *Twelfth Night* of dalliance and forgiveness (as he more or less puts it).[2]

The film has positioned the distant royalty of the Lordly queen in opposition to that of the tabloid persona in *Spy*, accepting neither as worthy of the human, one because too exalted, the other because too debased. One should be so lucky as to have a Dex to arrive at the penultimate moment and work wonders to free one of both. Failing that, one becomes a Marilyn, a Jackie, or a Diana. Except that one is never freed of the lower, the media spy, which is the film's other point.

That the media presence remains an ineradicable presence/nuisance in Tracy's life is made clear by the film's ending. Her wedding is—in spite of everything—captured on film by *Spy*'s own Sidney Kidd, who jumps out of the audience to snap photos in the middle of the service. A sequence of these snapshots ends the film. *Spy* has in this sense proved triumphant: you can't get away from the media, even in the early 1940s, when this film is made. This final sequence is shocking to the viewer as well as supremely funny. One can only assume these snapshots are in the magazine, whose pages are being turned at the end of the film. The film ends by turning into *Spy*, or so it might seem. That *Spy*'s photographic images end *The Philadelphia Story*, that they are in fact identical to film stills (of this film), with characters frozen in action, cannot help but invite the thought that (this) film is of the same sort as the photos in *Spy*. Or that it has somehow slummed its way down to the degraded state of the mag. How did they get in there, in the film? How did this happen? How did *he* get in there? Apparently Sydney Kidd has a way of slipping into any such occasion; he is an ineradicable presence in cinema—even that which seeks to keep him out in the private worlds of its upper-crust lyricism. Just as cinema gives birth to the queen and the star, so it aids and abets the tabloid, carrying forward its appeal into new heights: in spite of itself and because of itself.

I think *The Philadelphia Story* is saying that the infinite, unending moral and aesthetic task of film is to overcome the tabloid presence within itself, to make sure it does *not* reduce itself to *Spy*. This is no easy task, given that film and voyeurism go hand in hand and given that film reaches a mass audience (like *Spy*) and is part of the star system, the celebrity world. If the film rises above the peep show of the magazine, it is because it deals in the currency of stars, not merely celebs. And because it has a moral about the limits of those very beings, if anyone were to be so stunned by them as to wish to propose a life through their beaming presence. The film, one generation later, could be a warning against the cult of the star icon, who is also a goddess and stands on rooftops in a state of perfection, meaning perfect unhappiness. The film also reprimands those in the audience who, like Uncle Willie and Dinah, have just spent the better part of two hours spying on every character in this film, and from every possible position, meaning everybody watching the film including me. "I love it," Dinah says of *Spy* back at the stables, "It's got pictures of everything!" Like cinema itself, whose relationship with the cruder forms of scopophilia and scandal mongering are longstanding.

The Philadelphia Story is acknowledging its commiseration with the star system, the system of actress queens and larger-than-life beauties, not to mention the Hollywood publicity machine and its readers, with their desires for "pictures of everything." The art of cinema is a matter of discretion, implication, holding back, rising above, not avoiding or pretending to stand apart from it all. Whatever makes cinema art consists in this overcoming of debasement.

This is a film made at the inception of consumer culture and prophetic of it. Also its class consciousness allies it with the picture of British royalty, where Lord and spy converge. *Spy* has by now, more than fifty years after the play was written and the film made, found a way to take pictures of "everything," including Diana sunbathing with her paramours, sharing a personal moment with her young children, dying, including Jackie on Ari's yacht, including Grace sitting atop the mountains of Monaco in bleached elegance. The paparazzi took photos of Diana while she was dying, doing nothing to help, stole her soul as it passed from her body, considered her death nothing more than a whopping photo op. They were tried for this and for pressing on the Mercedes as it sped away from the Ritz on their motorbikes, found legally innocent. But at the moral level?

We have pictures of everything, we kitchen maids who saw it all through the keyhole. It is not simply that the tabloids have become more excellent at tracking their game and possibly more vicious in the way they kill. Cinema and the awe/worship of the cinema star or cult icon is itself a position of danger, illusion, voyeurism. The voyeur keeps the star at a distance, encages her as a queen, while the stalkarazzi hunt her. TV changes the entire way the hunt takes place. It is the other half of the equation linking public and star. TV is debased and also intimate, horrifying but also worthy in a way crucial to understanding the aesthetics of the star icon. I turn to it now.

seven

Teleaesthetics

I SHALL begin with the tabloid aspects of TV in full swing, with the talk show armed and dangerous, in other words with *The Birdcage* (1996, film directed by Mike Nichols,). "Today's contestants will be Yasir Arafat and Kate Moss." That is from the pen of Elaine May, screenwriter-comic of this masterpiece, a masterpiece concerned with two rightist turns in TV and tabloid: media-generated images of America written in the form of gelatinous one-liners, pseudo-nostalgic platitudes spoken by the senator from the new right, and genuine aggression (also endemic to the republican right, which is in perfect collusion with tabloid media). "We are in *Enquirer* Heaven," one of the journalists says to the other in this film, when they stumble on the fact that "Senator Keeley," leader of the "moral right," is in the house of a gay couple who own the drag club below. That the owner is Jewish, called Goldman rather than the more Protestant Coleman, adds spice to this media meal of fast food and fries, which it serves the ravenous American public along with ketchup and ranch dressing. Underbred, obnoxious harpies and overstuffed male half-wits these "gentlemen of the press," as Rosalind Russell would earlier ironize them in *His Girl Friday* (1940, directed by Howard Hawks). They will stop at nothing, leaving a trail of spilled coffee, parking tickets, and invective worthy of blood on the pavement.

The Birdcage is about reversal. The gay couple have a better marriage than their straight counterparts. The drag queen of this marriage, Starina (admirably played by Nathan Lane), is far more dedicated to family values than any of these self-posturing senators, newsmen, or media jerkoffs who drive their TV vans into handicapped zones and smirk at the moral indiscretion. Starina is a true American, wanting that simple Grovers Corners–*Our Town* life Senator Keeley prates on about but hardly pursues for himself.

The film is an enquirer inquiry into the vaunted obliviousness of the media, into its ruthlessness, its rough-and-tumble wrestling in the name of information circulation. Exaggeration? Compared to Fox News's recent attempt to publish a book by O. J. Simpson about how he would have committed the crime had he done it (which, the book will assure us, he did not), adding a TV interview to boot, one can hardly say yes. This reduction of news to business and news-business to entertainment, this further reduction of news-business-entertainment to guttersniping, and the ensuing dumb-down of America, has been a long time coming. Such news stations use telemorality like a blunt instrument and celebrate the cretin, the criminal, and the medicalized, which increasingly resembles everyone. They reduce complex realities to aggressively stated oversimplifications in the name of the American right and its right to fight (read: right to kill). Interviews increasingly resemble game shows, with each "contestant" vying to get their hand up first and shout out a one-sentence answer, which is all the time they will get unless they make prominent use of the word *tough* (tough on defense, tough of security, tough on health care, tough on baby's booties). If you're not tough, then you may be humiliated unless you allow yourself to be portrayed as an addict of something, a victim of some traumatic circumstance, a killer let loose in an Armani suit who didn't mean it and suffered an overdose of Diet Coke, a metrosexual liberal who has prayed his way back to manhood, a someone repenting on TV and telling all. This is particularly true of the celebrity, who having been hunted down and stigmatized as a drug addict or irresponsible childminder is allowed to preach confession as cure. Mel Gibson may be redeemed (or at least tolerated in absentia) because he publicly claims addiction to alcohol, Michael Douglas is, that is was until he was cured, a self-proclaimed sex addict, which justifies his infidelities, Winona Ryder a purveyor of grand theft and larceny, Richard Gere a Dalai

Lama clone who is said (by the media) to rely on squirming American rodents for his spiritual conquests. Ours used to be the age of
true grit turned true confession. Everyone has recently forgotten
they were victims of sexual abuse, parental scopophilia, an excess
of TV itself. The cure: more TV, dial a soap, where if you miss a
day's watching there is a 1–800 number to dial or one may go on-
line to find out what was missed about Pamela's abortion, Derek's
psychiatrist, or Melissa's biopsy. Welcome to the wonderful world
of illness and conflict, courtesy of your local pharmaceutical company with its anxiety-generating ads in the Sunday *New York Times*
about dreaded disease and miraculous medication. You no longer
need suffer from restless leg syndrome, medicine is available. No
wonder that so much has been written about the feminization of
America through television. We are all become soap opera queens
and pharmaceutical addicts.

   This is the TV news (worse because British) and the tabloids
(worse than that) that followed Diana around, stalking her, and
that a generation earlier tailed Jackie. They couldn't wait to get at
these women. Such aggressions have their sources in the spectacle
of the star system, the bigger the kill the better, be it (a hundred
years back) Stanley bagging Livingston or the stalkarazzi bagging
Diana. Of course not all TV is like this. The medium is as variable
as there is. I am a child of television. I lived on *Gilligan's Island*
with the Skipper too, the millionaire and his wife, I got poked in
the eye by *The Three Stooges*, I rode with the Cartwrights in their
big valley *Bonanza*, talked with *Mr. Ed* (who never talks unless he
has something to say), I baled hay with Eddie Albert in *Green Acres*
(farm livin' is the life for me . . . take Manhattan and give me the
countryside), I learned estrangement from *The Twilight Zone* and
black comedy from *Alfred Hitchcock Presents*. I perfected the art
of tying my shoelaces from Archie Bunker and the art of untying
them from *The Beverly Hillbillies*. I got my hairdo from It on *The
Adam's Family* (whose theme music starts like Beethoven's Fifth
but in the other direction), I received my first instruction in flirtation from the gals on *Petticoat Junction*. Often I wake to the jingles
of these shows, and if I happen to munch on a madeleine it is this

music that comes back. More recently, I have hanged tight waiting to discover whether Carmela will leave Tony on *The Sopranos* and I have swooned to the classy art deco interiors of Hercule Poirot's London. With its news and information, history and home shopping, interviews and talk shows, mysteries, comics, and music, TV is a vast medium, more a continent than an art form. In fact TV is exactly the size of America: Too much and not enough. I begin with the worst of the worst, since that is immediately pressing for the topic here, knowing there is much else worthy of discussion.

Given the worst of the worst, it is the *The Morton Downey Show* in the late 1980s that is to be singled out—and singled out as generative. This was a talk show on which contestants were encouraged to scream directly into each other's faces at close range while Morton Downey Jr. MCed like a boxing coach. *Big Time Wrestling* had been around for long enough to instill in the public an appreciation of staged spectacle—an appreciation of things known to be staged and enjoyed as such. Downey and his producers simply transposed this simulation of live action to the talk show where, as in the wrestling show, everyone knew people could get knocked around but no one expected to get hurt. This made for public spectacle in the Roman sense, the desire for blood and gore by a public jaded on its consumer perfection, bored with its own emptiness and anxious in the light of a chronically uncooperative world (which simply wouldn't do what they wanted). *The Morton Downey Show* encouraged people to attack each other like wolves without the real death sentence of the gladiator ever being passed (no thumbs down). Jerry Springer, former political Democrat turned rabbi-demagogue, around the same time began to foment rage: between mothers who slept with daughters' boyfriends or sisters who had wrecked the marriages of their siblings or deadbeat fathers who had disappeared without giving a cent to their children and now wanted them back because they'd become born again. Springer heated up the circumstance with audience questions of a particularly aggressive sort, calling on the "psychologist professional" at key punctuation points to provide a moral, then descending back to the white trash melodrama. The show usually ended with Jerry's homily about cooperation, sympathy, sharing, acceptance, and values: "Joan, you need to think about how to remain friendly to Bill without abusing him, and Bill, you can have your cake but not eat it too. Randee, if you deal with your mother's feelings this does not include biting, and Joan, lay off the sauce and think about who you really are. Remember life is about *give* and take, not just take. See ya all next week." At which point we cut to commercial break for automotive product or sanitary tampon, and Springer landed another ten million while Joan prepared to slice into Randee's throat offstage. Here we have the disease and the cure in one place, without any desire for resolution except as part of the act.

Talk shows used to be adventures in culture, conversations about art (Dick Cavett, David Susskind). However in the 1980s they began their long downward spiral and two principles began to reign: the license to kill and the therapeutic confessional staged before the gaze of millions. A long-lost dad would appear halfway through a program to embrace the daughter he had abused then abandoned while the mother discarded long ago shouted bloody murder. The American public drank deep of such scenarios, which then burst all boundaries by modulating to a new market: the news. The news, with Fox TV, became explicitly defined as a business, and a business running on entertainment values. Its idea of entertainment was staged assassination and pot luck humiliation; 911 simply fell into the lap of Fox TV; it was what the network was waiting for, because its aggression could be channeled toward a source everyone agreed to hate (was hateful). Osama made Rupert Murdoch a load of dough.

Talk radio became even worse: more virulent because invisible. And so, driving back from Chicago last November, passing through Western Michigan and tuned to one of the local AM radio stations, I heard a talk show host (he could have been no more than thirty years old, fresh from officer training school or the state penitentiary) say the following: "We recently heard that a group of professors from medical schools are petitioning congress to pass legislation limiting the ability of drug companies to advertise their drugs before the American public. Doctors from Stanford, Harvard, Duke, etc. . . . signed this petition guys. . . . Sounds elitist to me!" And that was that, the total story: "Sounds elitist to me." Apparently people listen, because this ridiculous little man on steroids has a radio show on every week; he holds a full-time job that is not professional big-time wrestler.

Oprah is better, more sympathetic, smarter, more genuinely about self-help, akin to Romance novels and group therapy sessions. "Physician: cure thyself," she said, and did it. Now she brings a concern transferred from her own past difficulties to the person interviewed and it raises the show above others. But the routines are nevertheless of stock character: suffering, victimization, weight loss and gain, marriage and divorce, what used to be called "Women's Programming" when radio and TV developed daytime soaps and women interviewing women in the 1940s and 1950s for stay-at-home moms, the out of work, and the elderly. Yes, we have all become feminized: weepy victims and consumers of passive aggressive rage.

What becomes of the star within this system, within this new media *culture?* In this media world all persons become equal, homogenized to a semigelatinous platitude. TV and radio homogenize through format, ideology, but also what P. David Marshall calls constant cutting: "Whereas the film celebrity maintains an aura of distinction, the television celebrity's aura of distinction is continually broken by the myriad messages and products that surround any television text."[1] No one has enough time to say anything. This system is tailor-made for the radical right, since it runs on reduction of real conversation and critical discourse to macho one-liners, a buzz of interruption resembling war, and a succession of bylines.

The star is usually reduced in this medium to celebrity product: packaged and formated between the poles of aggression, victimhood, therapy, and glowing adulation of every detail of his/her life. "Oh, you grew up on a farm, how wonderful! No running water, you are a SAINT to have gone so far in life. You had to put your mother in a home? How deeply scarred you must be! Your hobby is stamp collecting? You *are* an amazing, unusual man!" No one remains individual within the grip of this declarative system. The best stay away, live off-site (Meryl Streep in Connecticut, Johnny Depp in France). The worst laugh their way to the bank through it all. The half-wits believe in it, along with their producers and agents. Some collapse from the strain and have to be carted off to the Betty Ford clinic in designer ambulances. Some gain weight, then write a book about it. Many consume consumable substances in lieu of the caffeine they have given up on health grounds.

When television manages to grab a star and not debase her or him, it tends to exhibit an attitude of confusion or excessive deference, as if embarrassed that it is sitting in the presence of someone so exalted. This is the poor person's guide to greatness or what the kitchen maid saw through the keyhole. The authentic star (whatever that is) is something television cannot process, since talk TV's role is to homogenize and "celebrify." This is all highly relevant to the history of the aura, to Diana and her grace. For she was treated with kid gloves, being a special being of the kind TV does not usually encounter, and that throws it just slightly for a loop. At the same time, TV's confessional mode suited her perfectly.

"Compared with the film industry," P. David Marshall writes, "television has positioned its celebrities in a much different way. Whereas the film celebrity plays with aura through the construction of distance, the television celebrity is configured around conceptions of familiarity" (119). And so Diana was familiarized, further established as "people's princess" by going on TV, while also let off the hook with comparative ease: where they got her was in the news, with all those stalkarazzi.

TV also familiarizes, brings someone up close and personal, as if they've entered your living room for a chat. This is antithetical, in most cases, to aura formation, since it reduces distance and makes the protagonist more as opposed to less available to her public. Marshall says this:

> The familial feel of television and its celebrities is partially related to the domestication of entertainment technologies from the 1920s to the 1950s. Like radio, its precursor, television brought entertainment

into the home. And in terms of the common space of the family, the television occupied a privileged location in the living rooms of most homes in North America. The uses made of television were also modalized around its position as a family entertainment technology. The work of television production in the first two decades (from the late 1940s to the late 1960s) was the maintenance of a large mass audience, so that the same programs would be acceptable to all members of the family. Although different time periods were targeted by producers and advertisers for different audiences . . . there was a relative lack of audience differentiation beyond this level.

(119)

Television also familiarizes because it is produced so as to convey the illusion that it exists in live time. Indeed talk shows do, more or less, exist in live time, and, in the past, most weekly series were taped live in front of live audiences. The audience out there in television land, in the American home, has the experience of living with the TV personage when watching TV (Kitty Carlisle, Walter Cronkite), living alongside the TV character (Tony Soprano, Alexis Carrington), sharing their time zone, being there at the encounter between husband and wife, author and psychologist, at the Oprah show. We share a space with TV, or so it seems to us. When we sit down after work on a Wednesday, the TV personage is there, when the pizza arrives and the family sits down to a meal of TV dinner, the box faithfully delivers another segment of her ongoing story. This is the weekly talk show, the weekly episode: the span of lives lived in parallel.

When an audience watches talk show or news—even serial—the reigning sense is that we've the right to know all. Marshall puts this well about the talk show host:

In contrast to the film star, the television star who emerges as host and interpreter of the culture is treated as someone everyone has a right to know fully. His or her television life is subjected to daily scrutiny through his or her show. The talk-show host's opinions are obvious and forthright; his or her level of knowledge, humor, and

idiosyncrasies are his or her own. . . . Whereas the film celebrity proj-
ects a number of public images through performances as an actor,
the television talk-show host . . . is . . . constructed as clearly present-
ing him—or herself.   (131)

We get to know the talk show host in the manner of a weekly
encounter group, watching like a fly on the wall, unnoticed and
without undo involvement. This right to know is built into vast
segments of television: news, talk show, weekly encounter group,
and also soap opera and many sitcoms and serials. For the soap,
sitcom, and serial, we presume to know all about the characters,
who become surrogate persons with whose lives we continually,
rhythmically, intersect. Tony Soprano's character is deepened
over episodes to the point where I know him better than my own
cousins or aunts—know how he eats, what he eats, his idioms,
when he tends to blow up, what his nightmares are, his dreams,
how he treats his goumahs, his wife, his daughter and son, what
he likes about his house, his flirtatiousness, hostility, why he has
need of Dr. Melfi and how she is helping him. Possibly I know
more about him than I do about my younger brother. Yet we
shall never meet, except through the screen, where he acts and I
watch. When we get to the point where we can predict what the
character will do, what all of them will do, the series is exhausted
aesthetically—even if not yet shut down by the network it will be
running on empty. Knowledge is a goal, but fixated absorption
requires spontaneity, mystery. Between these poles a fictional TV
series rises and falls.

However caught up we the viewers may be in Tony's life, or
Oprah's guests, we watch from the quotidian safety of our invis-
ibility, our ability to switch channels if things happen we don't like,
our perpetual exit visa (we can shut off the TV). We remain glued
because we are hooked, and hooked in part because we are safe.
This feature of safety is shared with film, since the same kind of
apparatus mechanically generates both, one in which the image
on-screen is only there in two dimensions, outside the metaphysical ambience of our three, drawn from the past, even if only a

second ago in the case of "live" news. The apparatus may be similar in kind, but the aesthetic effect is profoundly different. Television speaks directly to us, or feels like it does this, film does not. This sense of mutual coexistence, between lives lived here in the world and there on the TV screen, is a relationship of secret sharer, mirror image. The viewer stands before the portals of a world that simulates ours but exists nowhere—call it a New Jersey of location and in the mind—and that proceeds by its own rules. This world, filmed in the past, is felt to be coextensive with our own in time. We do not experience it as a portal into the past in Kendall Walton's sense (even indirectly). The principle of coexistence, of *parallelism*, lends the TV series its sense of mystery within the ordinariness of familiarity. This is, if anything, the aura of television. However it is not an aura extending to the *characters*, whose lives are the stuff of intimacy with our own, running parallel and about which we claim to know all, which is part of the pleasure. They do not become stars in the film sense because they do not have the distance from viewers to solicit this way of imagining them. The series, the talk show, the news: these do not aim to generate aura around character (apart from the exception). Indeed TV seldom meditates on its own strangeness, for that very meditation is at odds with its fast-paced intimacy.

It takes film to stand back from TV and meditate on just how strange the series is, just how unearthly in mimicry, how bizarre in episode, how unfathomable in intensity, and seductive in claiming ongoing attention. This gaze by film on television is meditative but also a gaze of longing, envy, endless fascination. It is a gaze that seeks the absolute realization of the fantasy of TV, as if TV should aspire to the perfection of the movies. It is an unreal gaze, because TV has no interest in that, wants rather to blow the serial out of the water with something even better, a new serial, something that outrealities reality, turning the very stuff of it into a game show or whatever. Film's gaze on TV is that of an artist confident in his superiority, voyeuristic, omnipotent—finally, powerless. It is a gaze of fascination, arrogance, and also fear, because TV's aesthetics are so antithetical to film's aesthetics. This is not just about market

share, it is about how the same mechanism can produce antithetical results aesthetically. TV is film's Grace Kelly, always misbehav-
ing, wreaking havoc with its pretense.

What I am leading up to is that the star icon is among the few
to live a double life between these antithetical media in a way that
creates and completes her double existence, her star quality a figment of screen presence, that is, of cinema, her story the stuff of
television. The more we can learn about how these media oppose
each other aesthetically, the clearer the star icon's formation will
become.

The locus classicus of film meditating on TV is Peter Weir's *The
Truman Show*. Truman (Jim Carrey) is the adopted child of a TV
network; his life has been orchestrated by a TV director since its
inception. He does not realize that all the people in his life are TV
actors, all the sets TV studio environments. In order to keep him
"on the island," he has been given a story of his father's drowning
that has made him terrified of water, unable to cross the bridge
from this television island toward real life (away from this studio
world). The audience knows everything in Truman's life is fabri-
cated, that he exists in an enormous TV studio where everyone
but himself is an actor, that the actors have taken on the roles of
his wife, best friend, business associates, next-door neighbors for
the duration of the series or until they quit or, as in the case of the
one woman who breaks role and falls in love with him (and he her),
until she is fired. Truman is the only person who is unaware of the
artifice of his existence. But he feels this lack profoundly since he
spends his private hours planning escape (travel to Fiji), and pasting images together to replicate the visage of the woman who ex-
pressed genuine passion for him. He is like a man trying to recall
something that never was, but remains etched in his imagination.
She has generated in him an inchoate conviction that everything
else is cardboard—even his wife, who acts the role of someone
confused between love and shopping, life and lifestyle, love and,
dare one say it, television, and who has a massive capacity for hostility disguised by her plastic, California smile and white, bonded
Laura Linney teeth.

Truman is himself bonded, an indentured servant living in a world of dentures and prosthetic passions. The film is about his breaking free from this confined world of studio simulation where everything happens according to plan. If one can call it plan, since to keep this show alive without breaking through the artifice and exposing the reality fault lines to Truman requires a tremendous improvisation, an extreme virtuosity, on the part of the director, who is constantly cutting, pasting, sending characters here and there to ensure that the simulation remains life and Truman discovers nothing. He is a conductor, passionate, refined, orchestrating Truman's world, over which he reigns omnipotent and for which he has a father's love, surrogate of course, since Truman has been adopted by the network, not by he himself and is literally a child of television.

There has never been a better representation of naturalism prestidigitated through the genius and efficiency of the studio, of a studio island turned real in the way that malls, gated condo communities, and tropical resort hotels recast natural life in an unblemished, perfected format for those wanting a break from the real thing. This TV show satisfies the fantasy that life should become identical to its simulated double, and in a way that keeps the central character from recognition. It is our wish to become so mesmerized, so autonomic, so engulfed in what Cavell calls the automatisms of life. That is, to do this while also doing nothing but watching, to enjoy the fantasy of it, its voyeurism. The Truman Show is a San Diego of the soulless soul, a place of sun, regularity, and language always scripted, complete with an unending supply of happy endings. It is uncanny because it is a filament of life that cannot be touched and exists as if in the pristine form of Rousseau's natural man, nature being completely artifice in this case.

This controlled television environment is precisely what makes the TV serial a simulation. It is the extreme case of what every serial aspires to be, a combination of absolute regularity and complete spontaneity. The serial is characterized by restriction of domain. This is physical: the whole serial may be restricted to a taxi company, a network newsroom, a living room and kitchen, a small town in Alaska. But it is also essential to plot. Children come and go mostly on demand, or are absent altogether. There are no pets, no grandparents, no vacations are shown. No parents die, accidents rarely happen or happen with artificial regularity. The world of a TV serial is a world reduced to controlled circumstance, even if its interest derives from the well-placed accident, the ongoing chaotic uncertainty, the nagging doubt or problem. We crave oversimplification, finding it addictive, even if unreal. And *The Truman Show* responds to this fantasy by producing an entire life according to equally controlled circumstances. At the same time, because Truman inhabits this world not as actor but as person, it is his domain of freedom, a simulation of liberty itself, a mirror of life that doubles it in idealized form.

The price is Truman's emptiness, of which he has had enough for one lifetime after being given the gift of momentary passion by a woman who has seared his soul. He can hardly remember how she looked and is trying to assemble the memory by cutting out images from magazines, but her intensity has moved him forever. In a moment of magnificent TV melodrama (worthy of Barbara Cartland) he braves his fear of the water and sails the stormy television sea, facing and defying death from the artificial wind and wave the director has sadistically churned up against him so that he might die a noble TV death rather than break out of the the bug-under-glass television world in which he is contained. He reaches the other side, this Columbus in search of the new land, the free world. The film ends with his slamming into a wall, groping his way along its ridge until he finds a door that he enters. This door leads from his television world to the real one. As he opens it his lover, who has been watching him from the studio (which is where he enters upon opening this door), rushes down a set of stairs toward him. The film ends with his crossing through the door. He is no longer contained in the TV encomium. What will happen to him now that he is among the ordinary is no part of the film, for it is off television bounds. And so we do not see them meet, merely approach each other. Their life will be private, not at the center of television viewing. And so this "lifetime" show is over. The TV screen goes dead, nothing is broadcast anymore on this channel but pixels.

What is most interesting is watching those who watch. Throughout the film we have seen women in bars straining to catch the intensity of the moment, weeping for Truman, cheering him on, men fixed to the TV while in the bath, pounding at their bathwater like overwrought infants. We have seen a public of millions utterly engrossed in this real life teledrama. In the final scene of the movie one of a pair of viewers looks away from the empty pixels on screen and simply asks his buddy, "What else is there to watch?" And, after all that, this is all there is: what else is there to watch. Television land lived with Truman as if he were their uncanny brother for nearly thirty years, And we, the viewers of this film, lived with *The Truman Show* in a similar—albeit briefer—intensity. And now character, TV show, film are evaporated, disappeared into nothingness, as if this show never was, as if this life of Truman did not take place. All connection is severed; so much for memory, grief, attachments of any kind. When a series is over, it is as if it never happened to begin with. During its running we felt its parallelism to ourselves; once it is over it seems nothing but a distant mirror, now reflecting nothing except empty space. This is a film about the remarkable transience of the TV medium. It comes, it goes. It never was.

The other reason the film cuts off as Truman is about to be reunited with his lover is to make the film share something of the abrupt ending-in-process that pertains to the television series it is about. This sudden severing of audience absorption brings home how strange the presentness of this and all TV series are. For the TV series commands attention until the end, then disappears into thin air. Of course TV programs arise from others, through imitation and revision, reference and recollection. We remember them, remain attached to them, talk about them, laugh out loud about old Seinfeld episodes, find ourselves humming the theme music from *Gilligan's Island* with Gilligan, the skipper too, the millionaire and his wife, the professor and Mary Ann, here on Gilligan's isle. The series may become an object of nostalgia like anything else. But its temporality is that of a disappearing act. The series never seems to quite end, it is just pulled, or given a final plot, which seldom—like life—resolves all that was. There is no aesthetic conclusion to a TV series, it just goes away. And when it does, as often as not, we do not stand back and gape, but simply and immediately ask what else is there to watch. There always is something else: that is TV's plenitude. It is the moral of *The Truman Show*, a film that itself does not share in the evaporation of TV, but is a complete aesthetic object around which rescreenings may take place and a sense of finality may be found. Since there is no further goal to the series apart from articulating lives parallel to our own, with whatever twists (comedy, irony, violence, rich and famousness), the series is a floating opera whose purpose is not movement toward resolution, not Kant's "purposiveness without purpose," but Baba Ram Das's *be here now*, until . . . Art normally aspires to the symphonic, the formal integrity of music. It usually aims for formal completeness and final resolution of parts, whether pictorial, musical, or literary. The endings of series are almost always artificial, contrived better or worse, but rather like human deaths, things that just "happen." We know they are in fact pulled when they cease to make (enough) money or when a main actor dies or no one can stand to act the parts anymore or the writers quit. At the end of the sixth series of *The Sopranos* so many things could have been developed: the children and their ambivalence about the Mafia and their place as beneficiaries, Carmela and her ambivalence about being a Mafia wife, Tony's neurosis, the struggle between the various Mafia "families," and so on. This insufficiency of conclusion is the ongoing fascination of TV, which is like life as we live it in this aspect, but also unlike life in being an unending spectacle whose very point is to avoid resolution, to repeat elements of plot and character each week that in life would prove tiresome but on TV prove fascinating, to keep persons in a state of sameness rarely attained in life or, when attained, with paradoxical results.

This lack of "purposiveness," this failure of finality or refusal of it is a new kind of aesthetic structure in its own right, taken over

from radio and, perhaps, before that from serializations in magazines and oral culture (telling the same story again and again with
embellished variation). It makes TV strange—as if its image of life
is that we should be born and die without ever changing location
or developing as persons, as if we were characters freed from ultimate purpose, therefore from birth and death, indeed from the
inexorability of time. A character on TV seems never to age until
one day the character simply fails to appear, is killed off, or shown
to leave for Las Vegas or college or a trip around the world.

And so all endings to a series are unreal: the thing could go in a
hundred different directions like life. We the audience desire final
resolution—which *The Truman Show* does deliver with its ro-
mantically unreal moment of about-to-be union between Truman
and his fantasy lover—while also not wanting it, since we enjoy
the open-endedness of the series as a structured simulation of the
open-endedness of life. Any ending is unreal, just a way of allowing us to move on. This was brilliantly illustrated in David Chase's
final episode of *The Sopranos*. Audiences watched the final episode
after seven years glued to their TVs, expecting final resolution (so-
lution). Bookies took bets on what would happen. Some thought
Tony would get whacked, others that he'd turn FBI informer. Others that Carm would finally kick him out of the house. Still others that a war would destroy the entire Soprano "family," and yet
another group vowed that one of Tony's kids would get it, and he
would end in panic, as he began. The series had set up all of these
possibilities, since Tony was in the middle of a powerful gang war
and also chummy with the FBI guy formerly assigned to nailing
him (now transferred to the antiterror division).

So what happened? The final scene takes place in a diner where
Tony arrives first for a family dinner. We see him wait. Carm arrives second, ever the wife. His son Anthony Jr., now a producer's
assistant in the movies making trash, is third to arrive. We wait
with them; little happens. An Italian-looking man arrives and sits
at the counter. He stares over his shoulder at Tony two or three
times. Some eye contact is made, but Tony, relieved after a long
Mafia battle, seems not to notice. Is he failing to notice his killer?

How could he be so stupid, he is ever canny, ever the fox. Two African or Afro-American men arrive and play music at the jukebox. Are they accomplices or simply persons who have stumbled in at the wrong moment? And then it happens. His daughter Meadow, a lawyer and about to be married, arrives, but finds herself unable to parallel park. The Italian-looking man goes into the bathroom, aluding to that famous scene in *The Godfather* (1972, directed by Francis Ford Coppola) where Michael Corleone goes into the bathroom to return with a gun and shoot everyone at the table. Many viewers will draw the connection; Chase is hitting us over the head with it, wanting us to believe. Tension is now mounting; the scene is set for Tony's assassination. Meadow keeps trying to park and failing. Tension is now unbearable. Meadow finally manages to park and heads toward the restaurant. As she arrives at the door (without yet opening it) the TV goes blank, pure static, as if Chase were referring to the end of *The Truman Show* (perhaps he is?). Half of us watching across America think oh Christ we've lost reception and we're going to miss the ending! I almost turned off the set to reset my digital satellite connection. But then the final credits come on. We realize the show is over. Chase has played us like a violin, manipulated us into the obvious probabilities of endings and then refused to deliver any of them, anything at all that would provide the false sense of resolution, however unhappy or tragic. We wanted resolution (or at least a final narrative) so we could cry out, gasp, and then ask, "What else is there to watch?" Instead we are left hanging in Tony's life, not knowing if in the next half minute he'll be shot or perhaps be ready to defend himself, not knowing if he'll eat his supper quietly with family and then, the next morning or week or month, turn himself in, not knowing if he'll serve time, not knowing if Meadow alone will survive because of her delay, not knowing anything at all. . . . Time has stopped but La Vie Continue—continues beyond our knowledge. This final family supper (a Last Supper New Jersey style and not yet eaten), has equally been manipulated to intimate that everything has finally worked out in the family, another TV cliché that Chase invites and refuses to deliver—especially for this less than tranquil, self-deluded family. Given the impossibility of ending this—or any—series, Chase simply lets the TV go blank. Empty static is the death of the TV series, not of its protagonist. He's finally killed the series off while its characters, and we the audience, remain hanging, indecisive. Chase has killed not Tony, not his family, but *The Sopranos*, the TV show. It is literally dead.

This is perhaps the most honest ending television can invent, one that understands that, with the series, all endings are unreal, none persuasive beyond a certain point. The thing just stops as if it never was, evaporates into the thin air of pixels. The episode is called "Made in America" and is a meditation on television, both in Anthony Jr.'s entry into that unseemly world of producers and schlock and in Chase's setup for the ending.

Now to say we experience TV as an ongoing saga is to say each episode feels totally live, like it is happening right now. And when the series stops it feels like it has expired, died. Some have called this TV's *ontology of liveness*, a turn of phrase that really describes the phenomenology of TV viewing, our way of experiencing TV. A series is of course filmed in the past (before the moment of presentation), and it is filmed through the same kind of technology that makes a film. Its mechanical form of creation is the same. And yet it is experienced as live in the way film (usually) is not. Originally shows *were* taped before a live audience; those shows (as often as not directed by TV legend John Rich), *The Dick Van Dyke Show*, for example, have an improvised quality about them that high-production TV like *The Sopranos* does not, shot, edited, sound added. TV feels open-ended, film closed. TV is a lot closer to the way the present actually feels while we are in it than film is; TV's dimension is closer to life, since it always feels open to a future that has not yet been created for its characters (even if all the scripts have been written and the future is cast in dry ink). The sense of liveness especially pertains to news, of course, where the seven-second time delay quickly disappears from view and we've a sense of absolute simultaneity with what is being reported. For the news really is open, a slice taken from life as it is happening, however artificially cut and framed. Talk shows are the same: whether they were filmed seven seconds before you see them or seven years, they have the quality of incomplete adventures, where discussion may at any point take a nonscripted turn. Their scripted audience interventions are meant to augment the sense of liveness, of a show where "anything could happen" and does, even if the show is strictly controlled from the wings. Our sense of it is we don't know how it things will come out in the end (like our own lives) even though our pleasure (if one should call it that) in watching such shows depends on highly strict expectations about format (the "resolution" at the end, the shaking of hands, the sermon by the TV psychologist, etc.).

Television may, episode per episode, be as powerful, as articulately crafted, as well written, as high in production values, as film. On HBO it often is. Each particular instance of such series (*The Sopranos, The Wire*) is complete when watched, indeed cinematic and searing. But a question mark hangs over each episode, keeping us hooked. This desire to be present at the future portal of the show and then to see what happens in the next episode or newscast is TV's plenitude. The power of TV resides in its parallelism to life itself, in the intersection of the perfected power of the individual episode and the accretion of understanding about character and plot over the series. We live in the winding thread of its unfolding toward the future, always waiting. TV is perceived as a living, organically emerging thing. This quality of organic experience precludes a perception of aura around *The Sopranos* or any other series from forming, since these shows lack the requisite sense of pastness, lack the patina of something that is already disappeared from view and whose presence conceals the intimation of that absence. One seldom attaches an aura to one's own home until one is about to leave it. And the television show is a home that is ongoing, constantly being rebuilt.

Not all TV is like this, British crime dramas, taken from books, are not, but much of TV is, and for our purposes the kind that matters is.

At the basis of this organic quality is what Raymond Williams famously called "flow."[2] "A recent two-part commercial for Excedrin demonstrates the confident hold of flow, of continuity rather than rupture," Patricia Mellencamp wrote in 1990: "In part one, a man with a headache takes an aspirin, seemingly a conclusion; but after two ads and station promos, the commercial returns like a program, incorporating the time lapse into the sell and brief narrative: two minutes later the man's headache is gone."[3] Flow incorporates static. However, this flow is also constantly interrupted by commercial breaks, changes in series, flipping of channels, the making of snacks and ordering of fast foods. Our approach to TV is like wandering in and out of the river at will, we multitask, surf for different programs, get up and fix a drink, return to the TV with potato chips: but always the river, at least, until the particular series is pulled—which can happen at any time and inevitably does, generating mild anxiety, or what in her more extreme voice Mellencamp calls "catastrophe."[4] Such interruptions are contrary to the experience of an aura; there is no halo hovering over TV.

TV runs parallel to life, but is hardly the same. Were it identical to life—a complete mirror image of life—it would lose all distinctiveness. Instead it is a simulation of life whose difference consists not only in the fact of its fiction (enough said there!) but also in its degree of repetition. Repetition of plot and character elements allows us to form expectations about each episode before it happens, providing the comfort of Campbell's soup and sandwich, of fast foods where you know exactly what you will get. Carmela endlessly seeks escape from Tony but does not leave (or takes him back), she is always on the verge of rising above him but remains ever locked in assurance of her bourgeois life, dependent on his cash flow and his status as Boss: she will always live a lie. Mike Hammer gets whacked on the head with a truncheon every episode, but with each ensuing episode appears fresh as morning dew—with a cigarette hanging out of his mouth and a whopping hangover. Lucy will perpetually annoy Ricky Ricardo and get into a silent film mess: her response can be felt before it happens. Not that all elements are predictable; without uncertainty the sense of flow would cease to exist and with it the parallelism to life, the excitement of finding out what new twist will happen. And yet these characters are as far from time as animated cartoons, as regimented as paper dolls. This is the comfort of artifice, which in its complete value aims to take over Truman's whole life so that his spontaneity may be that of complete liberty while also completely constrained by formulae of expectation. It is in the double aspect of TV that desire and relief are generated, anxiety and its reduction. Tony always shacking up with a goumah/mother figure, never ready to leave her, Christopher always about to sink into a drug induced mire. Were most of these repetitive plot and formal devices vanquished our sense of TV continuity would be impaired. Without the expectation of repetition TV would lose its comfort zone and its addictive quality.

Without the exactitude of the setup, TV would become too much like life. This is the Truman element: same location (Lucy's living room, Tony's kitchen), same people (*The Office* crowd, *Gilligan's Island*, or *Green Acres*), same kinds of setup (Seinfeld always has the double plot, *Curb Your Enthusiasm* the comic humiliation). We love it when George in Seinfeld acts inimitably himself in ever crazier situations, when Eileen can be depended upon to say and do the nasty, when Kramer is yet again off the deep end, when Silvio juts his chin in the air in parodic agreement with his boss Tony. These things we depend on in a way we depend on a loved one, a cherished pet, a MacDonald's hamburger. As Warhol said, the key is to always eat exactly the same thing. It is then mother's milk, packaged in the form of a consumable. The pleasure we take in TV follows from its double life of spontaneity and repetition.

TV is a medium with multiple genres, and the point of importance for the star icon is that these genres over time become more and more alike. And so the news is formatted in sensationalizing forms that become referred in the public imagination to its genres of soap and serial. Diana's life, soap to the core, happens in news bites, which are felt to be the bites of a TV serial. So serialized, she ends up seeming fiction, while known as fact. The public imagination combines the episodic and already sensationalized story of Diana with the daily fare of programming. And so, what else does it have to watch besides *East Enders?* Diana! The Diana story, the Kennedy story, Princess Grace Kelly on the BBC become baptized in the waters of soap and serial and end up seeming real and simulation at once. Neither Jackie nor Diana can bear the stultifying embalmment this medium and the tabloids produce. Both try to flee, which makes the story even better television, even more soap-sudsy. TV participates in the making of melodrama for the star icon by straitjacketing her in its fictional forms.

The royals are already a TV show in gel, since they too live in a world of life scripted as ongoing and daily regimentation of form. Their artifice gives the public pleasure and makes them creatures of stiff upper lip, stiff whiskeys, and stiff emotions. Diana's royalty suited her for the TV completely, especially given her cinematic inability to endure it, her soap opera composure.

The end of the series is, David Chase tells us, a kind of catastrophe where flow is interrupted. Flow resumes by changing channels. But critical to the formation of the star icon is not merely the way her life becomes referred to the genres of TV (talk show, serial, etc.) but also the way TV responds to her catastrophe when it finally happens. Normally when daytime television is interrupted it happens like this: "We interrupt our normal programming to bring you . . . " and everyone in the audience cranes forward in their homes to hear which building was hit, which world leader assassinated. In *The Sopranos* catastrophe is reversed: it is because the screen goes blank that we *don't* hear any such news about Tony or anyone else, nothing finally happens apart from the death of the show itself. The first rule of TV at a moment of catastrophe ("We interrupt this broadcast") is to stay on the air at all costs. At the moment of the Kennedy assassination, the collapse of the World Trade Towers, the Princess Diana funeral television must remain fixed on the event, must never go blank. What Chase does is go blank before the catastrophe, thus making a catastrophe out of a catastrophe by not showing it.

This special condition of catastrophe is worth pursuing. Mary Ann Doane puts it thus:

> The catastrophe is crucial to television precisely because it . . . corroborates television's access to the momentary, the discontinuous,

the real. Catastrophe produces the illusion that the spectator is in direct contact with the anchorperson, who interrupts regular programming to demonstrate that it can indeed be done. . . . Television's greatest technological prowess is its ability to be there—both on the scene and in your living room. . . . The death associated with catastrophe ensures that television is felt as an immediate collision with the real in all its intractability—bodies in crisis, technology gone awry. . . . In fact, catastrophe could be said to be at one level a condensation of all the attributre and aspirations of "normal" television (immediacy, urgency, presence, discontinuity, the instantaneous, and hence forgettable).[5]

Because catastrophe breaks through ordinary flow, television's sense of liveness is strengthened by it. Distance is lost, deadness thrust aside, boredom replaced by powerful anxiety and fixation. We find ourselves driven into the intensity of the moment, as if the TV has become nothing but a telescope delivering reality straight—no chaser. What breaks through is the real, and with it our emotions of horror, grief, shock, anxiety, emotions of a directness not otherwise encountered in watching TV. For a moment TV and life have become one, the simulation has ceased, and the world is, traumatically, now inside the home.

This is one range of reaction, critical to public bonding around the media event. However, TV also trumps the real by assimilating the catastrophic event into its genres in a way that turns the real event into a weird simulation of itself and leaves the viewer dazed and disoriented. This happens through its refusal to stop even for an instant, its refusal to allow the viewer to change channels or go offscreen. Given the catastrophe, TV's goal quickly switches from ongoing reportage to avoiding the *deeper* catastrophe of shutting down and returning to ordinary programming (Chase's ending). Its mission becomes to continue the process of generating news around the event, of keeping it live at all cost. And so for the O.J. Simpson trial, which lasted an entire year, TV managed to stay on subject without surcease. When at last resort it found itself with nothing to show or say, when its endless dissections by expert and layperson of every detail of evidence ran dry, it would turn to ethics: "Do you think, George," one commentator would say to the other, "it is right for TV to be spending all this time on the trial when so much else is going on in the world?" To which George would reply, "Well, Dick, I see what you mean, but the public has a right to know, and anyway isn't it wonderful that we in TV land are able to raise this subject!" So much for what Andrew Ross calls tele-ethics, which is really a last-ditch effort to keep a media event alive.

By constantly hanging on the event, endlessly repeated, endlessly discussed, TV turns the event into an artifact of its fictional genres. The first time one sees the towers fall, Kennedy slump forward in his seat, the wrecked Diana Mercedes it seems unreal, deranged. The second and third time one sees this image (rapidly reprojected throughout the day), one begins to grasp the enormity of what took place in an instant. By the hundredth, two hundredth, five hundredth, thousandth time it is repeated, the image again seems unreal, as if it were an artifact of the screen rather than of life, as if it never really happened and its origins are in the TV box itself, endlessly replayed as if in some weird game. This inexorable movement from disbelief to realization to unreality is the mark of television's interference in trauma: for what it has finally done is refer reality to the repetitive formula of its *series*. The mark of a series is, as I have said, its endless instance of the same, either in setup or character (Lucy will always be fed up with Ricky's mocking, Gilligan will always try to get off the island and fail). But this sense of repetitive plenitude is convulsive when it happens over a catastrophe. The thousandth time the World Trade towers are shown falling they seem like they didn't fall at all except on TV, so much do they begin to approximate the TV series (again and again, again and again). It becomes like Carmela endlessly about to leave Tony, rather than the once-off trauma event it is. It is as if reality once happened long ago and is now living a second incarnation in screen form, which gradually obliterates the first. Our sense of the traumatic reality of the singular event goes haywire through excessive repetition: all sense of singularity is paradoxically lost. Our eyes begin to play tricks on our minds. The more we see the singular moment repeated the more we are inclined to read it as serialized fiction rather than one-of-a-kind fact.

These things are of the greatest importance for the star icon. Her daily melodrama happens in the news, but when her moment of catastrophe strikes it breaks through the bounds of ordinary TV by endlessly repeating, with all the derangement that comes about through such repetition. We cannot sort out what is fact and what is fiction about her life and aura, even though we know which is which. She seems to have been born and killed on TV rather than in life. The event is felt to be surreal. Finally the event becomes an addiction because the more we watch the more we need to watch to straighten out the deranged feeling we've gotten through too much watching! This is pure repetition compulsion, at which point TV has won out. We cannot exit the channel without withdrawal symptoms.

The star icon becomes what she/it is through her daily, weekly, or monthly presence on TV, but also, crucially, through the catastrophic event that only happens once and is endlessly repeated like the falling twin towers. Our deranged sense of her reality, which flows from the event, is something we will never get over. She will be caught in our television lens, meaning our imaginations, forever.

At the same time, the media event around her catastrophe has features of mourning, ritual, and grief that intensify its religious values. Dayan and Katz call this the aspect of "media event," which is "*interrupted* but *preplanned*."[6] Through the ritual of carriage, bier, church service, *Candle in the Wind*, catastrophe is mourned and also turned into what Dayan and Katz call "*reverence* and *ceremony*,"[7] the reverence of the ceremonial. These writers go so far as to say "*Media events blur the boundary between the sacred and the profane*."[8] And so trauma modulates into cult worship. Jackie may live in Greece, become the wife of the richest man in the world, sit on a yacht in designer sunglasses reading magazines, but she will always also be at the funeral in the public's perception of her, John-John beside her in his sailor suit, bravely bidding his father farewell. This is brutal, and among the few memory traces associated with the medium of TV, for wherever and whenever Jackie is shown, the public will project it onto what is seen, just as they will project Elton singing the candle at both ends of the Atlantic onto every image now shown of Princess Diana.

TV, meanwhile, quickly recovers from its own traumatic interruption; signals return and the ordinary fare of news, talk shows, and series come back on as if nothing had happened. Flow returns. This too can be shocking: to see Seinfeld back on where the day before it was all 9/11, knowing Jerry on TV is totally oblivious of the trauma that ripped the nation apart. One feels TV erases what it showed us the day before. This too sets the icon apart, confirms her strange uniqueness.

TV recovers but the star icon does not. She remains embalmed in the catastrophe, both through reruns and in the public's fixed imagination of her.

And so what a medium this is, this television land! How infinite in faculty, how noble in reason! The many conflicting signals of this medium are most graciously incanted in the final stanza of a poem by Lawrence Goldstein, a poem about growing up in an LA . . . LA home of the girl of the golden west, the thirty-four-inch television screen that, having intruded itself into the family home, leads to this:

> It is brighter inside than the glow of any tree.
> News briefs, reruns that kill an hour or two,
> Then game shows, a movie, and later news.
> Every four minutes merchandise bullies them.
> Mother and father, how shall I wield my love
> against the raucous cannibal of this house?[9]

Cannibal indeed, delivering fantasy and fact together, in its own manner, according to its own surrealism. Were one to want more history of this subject, this entrance of the cannibal into the home, this domestication, one would do well to consult the pages of Lynn Spigel's work: this scholar has written beautifully about its postwar entrance, with implications for family, gender, feminization of the audience, consumerism, and so on.[10] This homemaking feature is also crucial for public intimacy around the star icon, who is thereby in the home. All this is about domesticating her, while she also remains in that further galaxy beyond.

Does TV project an aura? Yes in the sense that a cult may form around this or that program, yes in the sense that the rerun may evoke the nostalgia of something gone by (the beautiful days when I watched *The Three Stooges* as a child in my mother's lap), no in the sense of its chronic refusal of things completed, things far away, things past. Film and television here part company with respect to the aura. While each can become a cult object (the *Rocky Horror Picture Show*, *Star Trek*), television, with its illusion of "liveness," its sense of daily occurrence, defeats the distance of the aura, the distance of the star. It becomes ordinary human life played out through ongoing story. The point is intimacy with these lives while they happen, fascination with them, as in a cartoon story (*Popeye*) serialized in accord with the endlessly repeated setup (newspaper office, marriage of opposites, Martian in the bourgeois household, Mindy with her Mork). A film is by contrast an object always already complete. Its pastness shows because it is not in formation. And its projection of distance is the opposite of TV intimacies and the audience claims to know TV characters in the manner of next-door neighbors.

Goldstein's poem captures the aura of TV, that strange, glowing incandescence, emanating from the center of the home, seen through the windows from outside. It is from outside that the familiar and domesticated cannibal in the home takes on the requisite distanciation, the requisite strangeness, for a halo or an aura. Watching its unearthly light pulsating against darkness and street lamp, seeing the inhabitants of the home cast in shadow, fixated on this box as if it were a religious icon, one can feel one has encountered a lunar landscape, a place of unearthly light and incomprehensible intensity. It is this luminescence that the artist Nam June Paik captured in his video installations when, for example, he arranged multiple TV monitors on the ceiling of a gallery in the Documenta show of 1987. Recapitulating the clerestory windows of the Gothic cathedral, signaling vast, inexplicable transfers of energy, these TV sets seemed, when drained of all content and reduced to the primordial incandescence of light itself, religious. The aura arises through the play of presence and distanciation, and this play comes in many aesthetic forms. It is as a rule contrary to the domesticated cannibal in the home and requires film (*The Truman Show*) or video art (Paik) to provide the right perspective for its formation.[11]

The film aura should not be idealized, even if I am in many ways nostalgic for it and if it delivers a star power the world is lesser without. TV is good to challenge its reign through its improvisational sassiness, its department store of plots and characters, and its powers of ordinariness. The film aura is, as often as not, the stuff of cult, star worship, voyeurism, misrecognition, and nostalgia—even if TV proves itself as often as not the boob tube. But I want to harp on this point: TV is antithetical to the films, and it is rare that a personage can sustain the aura of stardom while simultaneously living on the telly, rare that these antithetical media synergize that personage, rare that an icon can be created for and by the public through these diagonals. Only the star icon becomes what she is through her double life between cinema and TV/tabloid. She is queen (Tracy Lord on the silver screen) and talk show confessor and public intimate, about whom we claim to know all. Our desire to know and not to know but to remain star struck is the defining feature of our relation to her. But how, and in what ways, was Diana (among star icons the most televised, because the most recent) actually on this fifty-two inch box?

eight

Diana Haunted and Hunted on TV

AMONG STAR icons, Diana's mug was pasted in the TV news on a weekly if not daily basis in a way no other mugs were. The public saw its snippets of her cradling AIDS orphans, lounging in bikinis behind stone walls, shielding her face, every nerve ending strained in fury against the onslaught of the glitterati. The more she was hounded, the more obvious her contempt, the more the voracious British public wanted more. Their sympathy ever profound, they also felt themselves entitled to ongoing royal action: this was a prerogative of their sense of public ownership of the monarchy, of their sense that it is theirs (even if they can't join it). To paraphrase Groucho Marx, the only club I own is the one I can't enter because they won't have me: This is how I own it. The media taught the British public that with the cult around Queen Victoria and the public's demand for her return from seclusion after her beloved Albert's death. They wanted her to return from Scotland to Buckingham Palace, demanded it. Now the demand is for *visibility*, for media presence and then more media presence.

Apart from this daily fare of TV bite (Diana in Mozambique, Diana with Mother Teresa, Diana with paramour) Diana appeared only on marked, special occasions, which took on "iconic" value. Had she become a TV regular, her aura might have been eclipsed, with her turning into mere celebrity. However, the few occasions in which she figured for an extended period (as, for example, a guest on a talk show) were occasions that "coronated" her image. The word *coronation* is not fortuitous, for her wedding (July 29, 1981, St. Paul's Cathedral) was ranked (by her fans) as among the most memorable events in TV history. *Inside TV,* "the magazine for women who love television," put the wedding on its Top Ten Greatest TV Weddings of All Time list, which is surely a list composed in language that betters any talk show subtitle. More than seven hundred and fifty million viewers watched her arrive (a mere twenty years old then) in a glass coach, wearing an ivory-silk gown, with a twenty-five-foot train and ten thousand pearls. Charles wore his naval commander's uniform. It was something right out of Cinderella. America would have called it Camelot, British style, except that Camelot was British until the media draped the Kennedy clan in that mantle.

In this she followed the Queen's coronation, also avidly watched on television by nation and world. In the manner of such television specials the queen's aura was confirmed through her TV-broadcasted wedding, giving all viewers the sense of watching a secret, special event from which they would otherwise have been excluded. This is because the event was singular, a media event, not the usual fare of TV. And it was because everyone knew the queen was on TV but not of it, that TV was being graced to carry her marriage live. It provided the excitement of viewer intimacy and confirmation of her aura through the sacred ritual of the marriage. This is the critical point: Diana, like her mother-in-law, soon to be ex-mother-in-law, appeared in extended programs only fortuitously on the home box, even if it haunted and hunted her ceaselessly (although not as much as the tabloids, which were worse). Her appearances could remain special, thus preserving her aura while adding viewer intimacy by a public consumed with knowing all about her, while also with keeping her apart. This formula, always potentially unstable, would probably have collapsed had she lived, married Dodi Fayed, inherited his ersatz kingdom, become yet another (as it were) of Grace Kelly's Monaco brats or, worse, a production hostess in and out of television land (if that is the correct scenario).

The amazing thing is that Diana remained above TV even though she was of it in a way her mother-in-law was not, for she truly became soap opera queen (for a day) when she went on the talk show and told all. This transpired during her separation and divorce from Charles. At that point she had already become fair game, the object of endless speculation about her eating disorders, self-inflicted pain, and depressions. She was already stalked on-screen. In response she (and her publicists) took the decision to take charge of the box. She appeared only once in a talk show, which is critical, I think, to her retention of the aura of the screen (had she become a weekly sob story it would have gradually "degenerated," I suppose). Appearing once retained the singularity of the TV appearance, its halo of media event, although it was pure soap, pure confessional, with her being questioned about divorce, children, her need to hurt herself by a respectful and subdued interviewer and she answering in a mood of anger and quiet desperation. It was a case of woman tells all, and there is no one better suited to describe that singular appearance than Andrew Morton, her biographer. It happened during the divorce, after the appearance of the first edition of Andrew Morton's book, which Charles greeted as "the longest petition for divorce ever filed." And the TV appearance, wildly successful, led to a rapid conclusion of the proceedings. After that she was determined to strike out on her own, and the cameras followed: to Mozambique, Calcutta, etc. When they were not hounding her, stealing photos of her bathing in a bikini without the top, rebroadcasting the past affair with Captain James Hewitt (by then he was already author of an award-winning schlock volume on their affair), they were busy seeking out her private life, her nudity, tears, acts of charity and beneficence, her every moment with her children. She had understood herself to be the Marilyn Monroe of the next generation.

She had also understood herself to be an actor in a pantomime drama not of her making, even exchanging pleasantries about it with Jeremy Irons, who, when he said he was taking a year off from acting, got this response from her: So am I. And now the actress of pomp and circumstance became a talk show heroine. In the words of Andrew Morton:

While every other member of the royal family, most notoriously her husband, had used television to promote their causes and latterly to talk about their private lives, Diana knew that she would never be allowed that freedom by the Palace. She had enjoyed countless approaches from the world's most prominent broadcasters, including Barbara Walters and Oprah Winfrey, while in 1994 she was in detailed secret discussions about an ITV documentary of her life. In the end she reluctantly decided against co-operation, not only because Prince Charles was then working with Jonathan Dimbleby for his own programme, but also because of antagonism from couriers.

A year on, the increasingly beleaguered Princess decided to take matters into her own hands, secretly agreeing to be interviewed by Martin Bashir, a journalist attached to BBC flagship current-affairs programme, *Panorama*. . . . He soon realized that secrecy was essential if the project was to be a success. . . . Only by elaborate subterfuge would Bashir and his crew be able to record Diana's words. They used special cameras so as not to attract attention when they arrived at Kensington Palace on a quiet Sunday in early November, 1995. As a precaution Diana had dismissed her staff for the day, knowing that she could not trust a soul. . . .

This very British television coup . . . was a sensation. . . . The Princess, wearing striking black eye make-up, discussed her life, her children, her husband and her hopes for the future, with remarkable frankness. . . . She talked openly about her eating disorders, her depression, her cries for help, the enemy inside the Palace, and her husband's relationship with Camilla Parker-Bowles. In a phrase that pithily captured the problems of her relationship with Prince Charles she said: "There were three of us in this marriage so it was a bit crowded." At the same time she admitted her own infidelity with . . . Hewitt, who had previously told the story of their affair in a book. "Yes I adored him, yes I was in love with him," she said, adding that she had felt "absolutely devastated" by his betrayal when news of the book he had co-authored reached her ears. While casting doubts on her husband's fitness to rule . . . she spoke of her own ambitions not just for herself but for her children and the monarchy. "I would like to be a Queen in people's hearts . . ." The programme attracted the largest audience for any television documentary in broadcasting history.[1]

A victory for herself, and a coup for television, grabbing a story against the bulwark of monarchy in a subterfuge worthy of Mata Hari, the whole thing natural, appearing in what seemed live time, and yet stolen (taped) in secret from royal houses. For once she was hunted but in control of the hunt, even doing the hunting herself in the language of confessional, using her haunted state to her own advantage. There she is, telling all, held close before her public like a rare butterfly, ordinary in her extraordinariness, victim in her privilege, a royal in a gilded cage, wanting out, wanting freedom, so that for the first time (since Edward VII abdicated for love, anyway) the entire public could identify with royalty as conflict, drama, humanity, the whole public could see the far away up close as a mirror of their lives, as opening intimacy out to them. Her appearance was critical in turning the drama of the monarchy into a TV serial, since it reflected back on the whole story. Her public could take pride in her, sympathize. This was a blessing. And so, in this unique case, the persona was not demeaned by the talk show but instead the film star quality became grafted onto the forms of TV intimacy and serial drama. Film aura, royal halo, and TV talk show did what they so rarely do: they alchemized around her.

November 1995 is when this television program happened. This was less than three years before her death. She would have in those three years time to divorce, vacation with Hewitt before a jungle of cameras, fall in love with Dodi Al-Fayed, stay with him at the Ritz, consider forming a production company with him, then get it to the aorta in the speeding car, hound-toothed journalists madly photographing her as she lay dying in the auto wreck, never calling for emergency, never stopping the click click click to help. Dodi Fayed had told his father he would propose to her that night at dinner, probably he did. What her life would be now, with this absolute nonroyal from the family that owns the portion of England not owned by the Dukes of Bedford and Marlborough is unknown.

Would she have ended up making a film about the very paparazzi who instead "forced" her vehicle to speed out of control? Would she have become an ordinary TV presence, lost her aura? Would she have continued her charitable work? It is a story without an ending, like *The Sopranos*. What happened, happened, that is all, black screen, end of program, live drama of funeral. So the cult remains, transposed to the key of memory, dormant, gradually fading away, the star growing dimmer, since it no longer shines in front of the television camera or in public or in the tabloids, although the ten-year anniversary gave it momentary brilliance, and she retains millions of sad admirers.

One additional reason all mention of Dodi Fayed was "barred" from the Diana funeral was to keep form by refusing to dignify the miscegenation between royal and commoner, especially one from the Middle East with a darker skin. The condition of gifting this dead dame a royal funeral was that she was meant to remain royal in *sensu stricto*. Even Charles, Earl of Spencer, deleted reference to Fayed from the final paragraph of his penultimate draft. "And we thank Dodi Al Fayed for making her last weeks ones of happiness," is how the earl's text read before he expunged it.

What astonished and appalled the Queen (Elizabeth II) was that Diana was not simply out there in the media but a *star*, a figure of charisma in a system the Queen knew in her bones was more powerful than her own, monarchical "system." Diana was traitor to the throne not only by divorcing Charles, but by doing it as the reigning Queen of the Media, a version of Queen of the Night with its own lineage, a lineage not limned by history but by star quality, melodrama, and media/public interest. Diana's catapulting into this counterrealm, this enemy territory of royalty, this Tabloid Kingdom, happened not because Diana willed it. She was until the divorce disinclined and considered the media her enemy until the end. Rather, it crowned her in spite of herself. To be crowned by this system one does not have to like it: it thrives on the antagonism of its "royals," since this generates melodrama. When Diana, at the moment of divorce, actually chose to let the media into her palace and then appear on TV to tell all, this was her symbolic crowning as media queen, monarch in a realm of royalty that was proven decisively stronger than the world of the monarchy. This Elizabeth understood, which was why she then agreed to appear, one week after the death of Diana, on TV to speak "condolence and sadness." It was a way of putting up the white flag before a system stronger than her own. She did it to keep the monarchy intact.

nine

Star Aura in Consumer Society
(and Other Fatalities)

AND SO the rarity of a figure who lives a double life between film star royal and TV tabloid protagonist, a figure whose daily appearances on TV and whose single extended TV confessional interview exalt her aura and serialize her desperate story. She remains ever the star, while also the haggard talk show intimate in news broadcast, and it is as if once, and only once, around her life and persona, film and television alchemize. Aspects within TV, and between TV and film, that are usually antagonistic, synergize royally around her royal person to create her double persona. This is a moment when the media itself is employed around a single character to its fullest—and to extraordinary, peculiar, effect. One is stunned. (I am among the stunned.) This is the transcendental aspect of the icon: that the media seems to levitate around her. And since the public's only real way of knowing her is through this levitation, an absolutely peculiar aura is formed called the Diana effect, or the Jackie effect, the product of all these forces in combination with the public and its desires.

The star icon's rarity needs to be understood in terms of the unique synergy between disparate aspects of the media that happen around her: in terms of the media considered as an aesthetic *system*. This effect of the system is all the more rare in this our age of consumer capitalism, which reduces the aura of personages and things worldwide, one-dimensionalizing them. As the star and the diamantine reduce to the celebrity and the logo, the public secretly longs for that rare charismatic figure whose auratic values are not reduced but magnified. The desire for someone around whom to make a cult gets greater and greater. Then when it happens we are dazzled that our deep desire for someone beyond the mere one-dimensional celebrity can find fulfillment. It is not simply that the media have created a figure convulsing us like film stars do. It is that we are amazed she remains with transcendent power in the days of celebrity weather reporters and friends of Paris Hilton. Standing before Diana is like standing before a Michelangelo and realizing: once upon a time such a thing was possible to make. Except that no one made Diana, just as no one made Jackie or Marilyn or Grace, least of all these women themselves. A Diana is an accident of a system without intention or finality. And so she is all the more astonishing. It is a historical astonishment that a figure draped in the beyond can exist at all in our time. And yet, where there is homogenization there is the greater desire for such a figure, where there is marketing there is the greater desire for the above and beyond. And so the rare occasion becomes one in which the antagonistic features of the star system (film, tabloid, television) come into alignment. Earlier I called this alchemy. Now I shall call it aesthetic luck or aesthetic melodrama. Have it either way.

Consumer capitalism is context and also threat for the star icon. For consumer capitalism reduces aura to a marketing formula in a number of related ways. This happens partly through the mere fact of repetition. At first films were mysterious, the newness of the medium convulsing all manner of viewers, from surrealist poets to street cleaners. At first the *mere* fact of things and people appearing on-screen rang mysterious; all they had to do was exist in motion. However, as society became over time inundated with an excess of films, with an industry of such, the stakes of halo creation got harder and harder. Films and photographs gradually lost their power and had to work at retaining depth of aura, sense of mystery. See one cathedral and it overwhelms; see a hundred and only the best or most unusual retains this power. It is the same with films. Susan Sontag long ago made the point about the power of photography: see an image of a war victim and you are shocked, see a thousand and their power becomes derealized, their force rendered banal. It becomes a task of art, journalism, culture to keep the power in the image alive, the aura profound. Otherwise the image loses its force, fading like a worn coin.

This point has long been noted, above all by Andy Warhol, who adored repetition/reproduction precisely because of the deadening effect. A culture flooded with films and images of all kinds dulls experience and makes each liable to being treated as just another product in the market. This is the threat, I said earlier in this book, to the icon: the very circulation that creates and sustains her may also at any point deaden her. She is the living proof that aura may remain mysterious in spite of circulation, but the proof is also a fragile one. Especially (and this is the second point) because consumer capitalism thrives in a world of increasingly deadened images. It turns lost originals into new marketable logos in a marketplace where images are consumed with increasing rapidity (and vapidity). Genius, uniqueness, depth of aura: these qualities sell, but sell better as the item becomes one of a baker's dozen, a thing turned into a product type: the Picasso plate, the Marilyn in quadruplicate, *Rocky II, III,* and *IV*. Over time consumer capitalism has come to mine the peculiar distance, religiosity, and charm of the lost aura into mass-producible product formats. Aura is that which is retained after the thing is no longer in use, no longer the object of sacred activities. And so Ralph Lauren takes the rumpled hunting clothes of the mink and manure English country set and remakes them in a way that carries in the design a hundred British and Hollywood film images of "Old England." These can be purchased by the common man or woman, who wears them on the golf course, at the mall, to the movies. The aura of Old England becomes an advertising logo, a suit of clothes tailor-made for lawyers and stockbrokers on the rise, as in Julian Barnes's novel *England, England.* Before you know it, six-year-old daughters of real estate magnates dress in Marie Antoinette wear, infants in Edwardian finery. The *New York Times Travel* section is mostly an occasion to advertise fashion shoots in faraway locations (blonde models in white on white Anna Karenina furs posed in the tundra of Lapland, Indian models in salwar chemises reclining on the parapets of Lake Palace, Udaipur, their jet-black hair overhanging the waters awaiting a golden prince to climb up and rescue them). In the Ralph Lauren store you may purchase a royal bed embossed with the House of Lauren logo with matching ensemble for your purebred dog: you too are a duke or earl strutting your stuff on the streets of Manhattan or Newport Beach, California.

The celebrity values in the world of modern art have turned logo in exactly the same way. I talked about this earlier: about the Picasso item, plate after plate, bowl after bowl, each reeking of the name PICASSO, so that the cardiologist or lawyer who purchased would be assured immediate recognition value for self and friends.[1] Between the logo and the celebrity Manhattan bowed down before its bad boy artists of the 1980s, whose "megagenius," loudly proclaimed by gallery, critic, artist, and collector in one shrill voice, led to a long line of would-be purchasers standing in line for their share of product, resembling those who call months if not years in advance to book at the trendiest restaurants and still arrive to an hour's wait at the bar. If you wanted a Julian Schnabel painting you put your name down on the list, waited months while he dished out the stuff, and then proclaimed joy and rapture when your name finally came to the top of the list and you got the call: "Your Schnabel is ready." It was like an opening of the heavens. "Is it rare?" you might have asked. "Over easy?" "I like mine with sun dried tomatoes and fresh basil." "It measures thirty-six by seventy-five, is a picture of a dead cowboy riding the empire state bldg like a Ford bronco and as a special treat the canvas is dripping with the entrails of dessicated sheep," the answer came back from the gallery director over the telephone. "Does it have red in it, I like a bit of red in mine," you asked. "What does it matter?" came the response, "It's a Schnabel, and by the way the meter's running out on this call, please make up your mind, there are plenty of other people who will take this work in a second." So much for the rapt contemplation of the art object that characterized the days of awe, that is, aura, the days of aestheticism, the nineteenth and early twentieth centuries, when persons took folding chairs into churches and museums to protect themselves from backache while the hours passed as they practiced brushwork. In the New York of the 1980s you ended up buying sight unseen, brand recognition was enough.

This turning of the aura into one-dimensional marketing logo with brand/celebrity recognition happened most forcefully in film. The starring role became reduced to a product type or star logo. With the advent of high-concept film in the 1980s, Demi Moore was marketed as an army private, hooker, working girl, brain surgeon, nuclear physicist, evil genius. Each role became reduced to the Demi-god, that false logo with sneer and sulk in tight jeans. The type became one-dimensional, the character reduced to the one-dimensional type. Film turned its own star auras into high-concept marketing images with instant recognition value and market penetration. This increasingly became the essence of film, which is why stars can have such enormous control today, with the power to end up owning their own production companies and directing their own films (with often miserable results): this while also adopting third world babies over the Internet. Is this the world of *Will Success Spoil Rock Hunter* (1957, directed by Frank Tashlin)? You bet! Rita Marlowe (Jayne Mansfield) is now the "titular head" of her own production company, orchestrated by one Rockwell Hunter (Tony Randall), aka Lover Doll, selling stay-put lipstick stay put put, stay put stay put stay put, to the tune of "Old MacDonald Had a Farm," just like in the movie.

In the 1990s the public finally became sick of this scene, preferring an iota of mystery and romance in their films and film stars, and Hollywood again heightened up, readmitting some peculiarity/individuality in their films and stars. Contemporary cinema veers between these poles today. Thus is Benjamin half correct. The aura has withered and is in constant danger of withering more. But not because mechanical reproducibility dispenses with it per se. Rather because mechanical reproducibility recreates it but it then becomes mined by consumer capitalism, which degenerates it (the magic of film) into a glamour component, an advertising allure surrounding people and things on-screen.

This process has proved crucial to the market expansion of the star system. More newspapers got into the act; more independent producers wanted their piece of the action. There was more distribution to be had, bigger pictures to be made, and more money. The value of the star depended on her rarity, and the studios always understood that (keeping her shielded, while also making her available in limited quantity). But with newsprint and, later, television, *more* of the star meant better copy and better sales. Recall that quote from *People* editor Richard Stolley: "We're scouring every facet of American life for stars. We haven't changed the concept of the magazine. We're just expanding the concept of 'star.'"[2]

With the expansion came the increasing need to divorce them from values of talent, natural-born beauty, apple pie values. Television allowed for more rapid turnover of stars (a new series, talk show host, confidante every year). As in all situations where increasingly rapid turnover of product requires increasingly rapid branding of product, the public then was given less time to bask in the aura of the star, who mostly did not have it anyway. Rather she should be quickly consumed and discarded. And I mean *she*, not he, because male stars still continue on with aura. Sean Connery can still (as it were) James Bond his way through a film at the age of seventy plus, bedding down with partners the age of his great-grandchildren. Clint Eastwood can still stride his way through all manner of battle. Jack Nicholson has yet to reach his peak, become as good as it gets. But actresses (e.g., those of the woman persuasion) flutter in and out with the speed of last year's ready-to-wear, which is what they've become by the age of forty plus. Females are simply more commodified than their male counterparts.

As the market speeds up, more time needs to be saved. Rapid turnover of product means increasing conformity of product: there is no time to invent the new, rather the market depends on transfer of logo from old to new product every season. Each star becomes branded with a slight but distinguishable variant from the last, like a new ready-to-wear line or toothpaste. Stars appear and disappear with the rhythm of new and outdated fashion. Each stands for the new, but marketed as a variant of the old. This simulacrum of newness was perfectly illustrated by the sign that was plastered onto the curved wall of the Beverly Center, Los Angeles, in the early 1990s, around the corner from where I happily lived in the early 1990s: "Don't blend in/The Beverly Center," the sign read. Don't blend in, shop: shop for the particular color and size that appeals to you. Stand out by purchasing a designer's particular make of blended fabric that everyone else purchases to (similarly) not blend in. March to your own drummer in accord with the designer's formula for individuality. Think you are different while taking silent comfort in being the same as everyone else who lives through the designer's mirror. Pretend, therefore, you are Henry David Thoreau while acting like a Ralph Lauren model. Over time (call it the twentieth century), capitalism comes to mine this public consciousness of images in the form of product values, *image types* containing the peculiar distance, religiosity, and charm of the aura but in mass-producible/ownable form. And so Ralph Lauren and his mink and manure English country day world remade in New York with the aura of Old England as its advertising logo.

The star system expanded into most every aspect of public and private life (with it privacy became a public matter, a matter of public advertising, confession, talk show, and therapy). Broader changes in capitalism from industrial production to information, communication, and marketing systems led to generalization of star quality into market quality and creation of this star market quality in every field of business, from politics to religion. The extrapolation of the star system to American politics of course began with the Kennedy presidential debates, in which Nixon, it is said, lost because of his five o'clock shadow, that edge-of-night glow of beard that cast him in a sinister aspect. Nowadays presidential debates are a matter of physiognomy, gesture, and synced language—opinion/fact/analysis/vision and sound editing are all too often secondary. The Reagan principle is now permanent: always look sincere, always sound profound. Carry the movie version of history behind you like a terrible swift sword. Don't say anything too controversial, because even if you do the media won't pick it up except to nail you. Reduce your ideas to marketing slogans so you can brand yourself vis-à-vis other candidates in ten second sound bites. Don't try to develop any complex ideas, you'll never have the time, and, anyway, no one will listen. When these conditions of branding are generalized to all aspects of American public life, we've got headaches, and not only political. Film itself suffers, since its artistic uniqueness gets understood as a mere variant on media culture generally, a brand among others, like shampoo. Such migration of the aura away from the silver screen into the hair salon, the department store, the presidential debate is exactly what Warhol reveled in, exactly what turns the aura of film into simulacrum, a whiff of the aura in the form of marketing value. It is making things more difficult for the icon to shine through the maze of getting and spending, the marketing muck.

If I sound polemical, it is not because I think the star icon and her aura without its own historical problems. Indeed all art has its problems: Renaissance art, in celebrating violence, self flagellation, not to mention the abrogation of sex, is not exactly value perfect either. What we speak of in adulating the icon is the perfectibility of forms, a dimension of transcendence now, with the marketing of the aura, entirely buried. One wants aura because of its power of absorption, its capacity for glow, its relation to art. One wants it so one can also criticize it. Without it, there is no chance for art today—not much, anyway, given the rule of the market. One of my favorite authors on art is Robert Hughes, whose lambasting of the times I find appealing, even if I am not in exact agreement with his (many) opinions. His move, often exhibited in print, of reverting to past art as a bulwark against what he dislikes in the present is not a move of pure adulation of the past—not at all. It is a move that recruits history to criticize the present, to gain perspective on it. That is how I would like to use the aura here.

That and to speak of how rare the magic is when it happens in this system, in the guise of the star icon, even if she is trapped, and we are cruel in our way of watching her. She is a route into the tenor of the times, a song often off-key but also containing the allure of Ulysses' sirens.

Even the royals have begun to market themselves through the media they also despise, a media that continues to make them miserable. Recently two journalists were arrested for having wiretapped Charles's and Camilla's phones, in the hope of more juicy tidbits of the "I'd like to be your knickers" variety—language that by any standard other than that of purebred dogs is not exactly hot. The BBC released a suite of four DVDs in 2005 of short documentaries that had been broadcast on their channels: *King Charles and Queen Camilla: Into the Unknown, Princess Camilla: Winner Takes All! Prince William and Prince Harry: Prisoners of Celebrity*, and *Harry: The Mysterious Prince*. Are we talking about Harry Potter here or what? These titles are themselves a game of Quiddich with the market by those who, simulating themselves, seek marketable images of what they once were. The DVDs would be better were they filmed by the makers of the Harry Potter series, which also mines the elite English past for castle and lord. Schlock of course, simulation certainly, late capitalist marketing product indeed, but the worst thing in life is not being talked about, not being *out there* on your advertising terms, when the world is everyday tabloiding you on its terms. And they fail because the other royals simply don't cut it on the level of star quality, resembling bull and heifer rather than classical beauty, lacking entirely in physiognomic response. They are a smirking, silly group and that is how they appear. A media is friendly only to the figure it causes to glow. And so the failure is vast: the royals are long on lineage and short on persona.

But those who film them know this. Their point is not to make Harry a star but to erase Diana from the royal picture, to show the royals can have a life on television without her to break the public's ironclad association of Diana with the media. The royals now understand that with their falling ratings even they need an aura cast by the media. Since they've little to offer, they've ended up reducing the royal aura to a one-dimensional marketing logo of the *People* /Ralph Lauren variety. There are no more blinds on the windows of Buckingham Palace. No one can exit the media today. So much for the aura of royalty, however long the royals may or may not last.

And so in the half-century during which the icon arose, from Marilyn (her funeral did not take place as a TV media event and she was not on TV regularly) to Jackie (who was) and Grace (who was intermittently) and then crowned with Diana's starring role in all media, the threat to the icon has increased along with, paradoxically, her media presence. Her survival value is all the more precarious, while the need for her is greater.

Or is it? Is the need greater? Maybe these days we prefer our idols to be contestants on *American Idol*? Or stars on reality TV programs like *Survivor* and (my favorite) *Celebrity Rehab*. Maybe the star icon with her halo for a crown is a thing of the past? And, if so, would this be a good thing or bad? Hard to say: some long for her type, revel in it, find it astonishing, but the "reconciliations" it offers between beauty and suffering are false. And the cult around her conceals blood lust for her pain under explicit admiration for her stardom. It is a whopping paean to voyeurism in all its aspects. On the other hand, in the absence of the star icon we are left with the aesthetics of Demi-gods and Handsome Harrys, of *People* and *American Idol*. These aesthetics preserve none of the mystery of a Garbo, Kelly, Diana, in spite of the royal DVD's optimistic titles.

And even if they did, how terrible are the ways in which star icons, with all their aesthetics, get narrated day by day? How terrible are the ways their very publics swarm toward them like moths before the flame? What kind of mass confinement is this, and how can any human being bear it? So we might ask: where can human mystery find better presentation for star icon and celebrity alike? And we might ask: where can actual lives of human suffering and human beauty better be pictured or narrated for mass audiences? The paradox is this. The very media that are called on to narrate these celebrity and star lives levitate them, hunt them, curdle them, and turn them into the stuff of silver screen and soap opera. The pressure on the media to continue to tell their stories in a celebrity voice proves overwhelming. Other kinds of lives can sometimes fare better on the media: celebrity and star icon do not. And the tendency is for the media to celebrify, to turn ordinary lives into celebrity lives. Can the media reverse gears? Will the public be disappointed?

This is a question about the state of the media and also about the state of public understanding and public desire. It is a question not only about the narration of celebrities and star icons, but about lawyers, physicians, out-of-work employees, United Nations officials, and American presidential candidates. It is a question about the flow of information through the media, a media that is increasingly identical to the public sphere. It is a question about how the public seeks to understand this flow of information and the way public desires are shaped and satisfied by this flow. Such issues of truth, allure, and illusion weighed on intellectuals throughout the twentieth century. They are even more pressing today.

But here I am still stunned by the icon, still overwhelmed by her aesthetic power, still fascinated by her terrible story. Perhaps I am more like the weeping congregants at the Kennedy funeral, the Garland concert, the *Candle in the Wind* than I think. Should I go into celebrity detox? Or is the star as icon a kind of art that one should not, cannot relinquish? Even now I find myself of two minds on the subject.

Notes

1. The Candle in the Wind

1. *BBC World News*, live coverage of Princess Diana Funeral, September 6, 1997.
2. Erwin Panofsky, "Style and Medium in the Moving Pictures," rewritten 1947, in Leo Braudy and Marshall Cohen, eds., *Film Theory and Criticism* (Oxford: Oxford and London, 1999), pp. 279–292.
3. Ibid.
4. Ibid.
5. Robert Turnock, *Interpreting Diana: Television Audiences and the Death of a Princess* (London: British Film Institute, 2002), p. 41.
6. Richard Dyer, *Stars* (London: British Film Institute, 2004), p. 125.
7. Rosalind Coward, *From Diana: The Portrait*, forward by Nelson Mandela (Kansas City: McMeel, 2004).
8. Richard Johnson, "Exemplary Differences: Mourning (and Not Mourning) a Princess," in A. Kear and D. Steinberg, eds., *Mourning Diana: Nation, Culture, and the Performance of Grief* (London: Routledge, 1999), p. 37.
9. See for example Kevin Noa, *Two Princesses: The Triumphs and Trials of Grace Kelly and Diana Spencer* (First Books, 2002).
10. P. David Marshall, *Celebrity and Power: Fame in Contemporary Culture* (Minneapolis: University of Minnesota Press, 1997), pp. 81–82.
11. Robert Musil, *The Man Without Qualities*, trans. E. Wilkins and E. Kaiser (London: Picador, 1979), chapter 13.
12. David Gritten, *Fame: Stripping Celebrity Bare* (London: Allen Lane, 2002), p. 7. I wish to thank the Los Angeles entertainment lawyer Michael Perlstein for bringing this superb book to my attention.
13. Richard Stolley quoted in Joshua Gamson, *Claims to Fame: Celebrity in Contemporary America* (Berkeley: University of California Press, 1994), p. 43.
14. Ibid.
15. Leo Braudy has richly detailed the history of fame in the West. See Leo Braudy, *The Frenzy of Renown: Fame and Its History* (New York: Vintage, 1997).

16. David Lubin, *Shooting Kennedy: JFK and the Culture of Images* (Berkeley: University of California Press, 2003). In that book Lubin speaks to the culture of movies that provides context for the public's projection of film qualities onto this pair.

17. Classic papers in this regard are Laura Mulvey, "Visual Pleasure and Narrative Cinema," *Screen,* vol. 16 (1975); Linda Williams, "When the Woman Looks," in Mary Ann Doane, Patricia Mellencamp, and Linda Williams, eds., *Revision: Essays in Feminist Film Criticism* (Frederick, MD: University Publications of America, 1984); and Kaja Silverman, "Lost Objects and Mistaken Subjects: Film Theory's Structuring Lack," *Wide Angle* vol. 7, nos. 1–2 (1985).

18. The persona's relation to opera is profound. For a discussion of the feminine role in grand opera, see, above all, Catherine Clement, *Opera: The Undoing of Women* (Minneapolis: University of Minnesota Press, 1988).

2. There Is Only One Star Icon (Except in a Warhol Picture)

1. Tina Brown, *The Diana Chronicles* (Doubleday: New York, 2007).

2. Wendy Leigh, *True Grace: The Life and Times of an American Princess* (New York: Thomas Dunne, 2007), p. xi.

3. Leo Braudy, *The Frenzy of Renown: Fame and Its History* (New York: Vintage, 1997), p. 9.

4. Ibid., pp. 580–581.

5. Ibid., p. 581.

6. Ibid., p. 6.

7. Andy Warhol and Pat Hackett, *POPism: The Warhol Sixties* (New York: Harcourt, Brace, Jovanovich, 1980), pp. 39–40.

8. For a superb introduction to this desultory state of affairs, mostly concerned with Picasso's celebrity, see John Berger, *The Success and Failure of Picasso* (New York: Pantheon, 1989).

9. Philip Fisher, *Making and Effacing Art: Modern American Art in a Culture of Museums* (Cambridge: Harvard University Press, 1991).

4. A Star Is Born

1. David Gritten, *Fame: Stripping Celebrity Bare* (London: Allen Lane, 2002), p. 16.

2. Ibid.

3. Ibid, p. 18.

4. Joshua Gamson, *Claims to Fame: Celebrity in Contemporary America* (Berkeley: University of California Press, 1994), p. 20; his quote is from Neil Postman, *Amusing Ourselves to Death: Public Discourse in the Age of Show Business* (New York: Penguin, 1985), p. 141.

5. Gamson, *Claims to Fame,* p. 24.

6. Ibid., p. 25.
7. Quoted from Maureen Dowd, "Mel's Tequila Sunrise," *New York Times*, August 2, 2006.
8. Ibid.

5. The Film Aura: An Intermediate Case

1. Cf. David Freedberg, *The Power of Images: Studies in the History and Theory of Response* (Chicago: University of Chicago Press, 1989).
2. Walter Benjamin, "The Work of Art in an Age of Mechanical Reproducibility," in Leo Braudy and Marshall Cohen, eds., *Film Theory and Criticism* (New York: Oxford University Press, 2004), pp. 791–811.
3. Which is why he praises Brecht's "epic theater": for refusing theater per se and the aura of the actors associated with it.
4. For a good discussion of this diminished role of the aura in Benjamin, see Miriam Bratu Hansen's introduction to Siegfried Kracauer's *Theory of Film: The Redemption of Physical Reality* (Princeton: Princeton University Press, 1997).
5. David S. Ferris, ed., *Walter Benjamin: Theoretical Questions* (Stanford: Stanford University Press, 1996), p. 45.
6. Erwin Panofsky, "Style and Medium in the Moving Pictures" (1947 [1934]), reprinted in Leo Braudy and Marshall Cohen, eds., *Film Theory and Criticism*, 6th ed. (New York: Oxford University Press, 2004), pp. 289–302.
7. Ibid., p. 302.
8. Panofsky reduced the screenplay to something that could be whatever it wanted so long as it did not dominate visual effects. This he called "the principle of co-expressibility." Where dialogue is in danger of dominating, something visual happens to match its importance, often a close-up. Panofsky's particular way of thinking this through does not do justice to the screenplay, which is not the "junior partner" in a movie, kept in check visually in case it should assert itself too much. The meanings found in the screenplay are generative for everything that happens. Coexpressibility is right insofar as the movie must never become too "talky." However, a better model for thinking the relationship between screenplay and camera is one of mutual generativity within a complex system. The screenplay is written—sometimes adapted from a documentary, work of history, short story, novel, play—with visual realization in mind. Flow, rhythm, character, plot are all imagined for the screen, often with particular actors, actresses, and locations. Equally important, a sound film synergizes visual rhythm with sound rhythm; sound becomes central to physiognomy. The medium of sound film is not simply visual reality as such; it is also sound.
9. Stanley Cavell, *The World Viewed* (Cambridge: Harvard University Press, 1979), p. 23.
10. Kendall Walton, "Transparent Pictures", *Critical Inquiry* 2.2 (1984): 251.

11. Kendall Walton, "Pictures and Photographs: Objections Answered," in Allen Richard and Murray Smith, eds., *Film Theory and Philosophy* (New York: Oxford University Press, 2003), p. 60. Walton's original essay on this subject was "Transparent Pictures" (see note 10).
12. Walton, "Transparent Pictures," p. 253.
13. In formulating this I am hugely helped by my student, Everett Kramer. We hit on this idea more or less together and at the same time, but he was just perhaps a little clearer. His help has been invaluable to me.
14. Ludwig Wittgenstein, *Philosophical Investigations*, no. 122, trans. Elizabeth Anscombe (New York: Macmillan, 1958).

6. Stargazing and Spying

1. Tania Modleski, "The Master's Dollhouse: *Rear Window*, from *The Women Who Knew Too Much*," reprinted in Leo Braudy and Marshall Cohen, eds., *Film Theory and Criticism* (Oxford: Oxford University Press, 2004), pp. 849–861.
2. Stanley Cavell, *Pursuits of Happiness: Hollywood Comedies of Remarriage* (Cambridge: Harvard University Press, 1981).

7. Teleaesthetics

1. P. David Marshall, *Celebrity and Power: Fame in Contemporary Culture* (Minneapolis: University of Minnesota Press, 1997), p. 121. Further references will appear parenthetically in text.
2. Raymond Williams, *Television, Technology, and Cultural Form* (London: Routledge, 1990).
3. Patricia Mellencamp, "TV Time and Catastrophe: Or Beyond the Pleasure Principle," in Patricia Mellencamp, *Logics of Television: Essays in Cultural Criticism* (London and Indiana: British Film Institute/Indiana University Press, 1990), p. 241.
4. Ibid.
5. Mary Ann Doane, Patricia Mellencamp, and Linda Williams, eds., *Revision: Essays in Feminist Film Criticism* (Frederick, MD: University Publications of America, 1984), p. 238.
6. Daniel Dayan and Elihu Katz, *Media Events: The Live Broadcasting of History* (Cambridge: Harvard University Press, 1992), p. 7.
7. Ibid.
8. Ibid., p. 207.
9. Lawrence Goldstein, "Homesick in Los Angeles," originally published in *Poetry*, vol. 73 (May 1985), reprinted in Lawrence Goldstein, *The American Poet at the Movies* (Ann Arbor: University of Michigan Press, 1993), pp. 249–250.
10. Lynn Spigel, *Make Room for TV: Television and the Family Ideal in Postwar America* (Chicago: University of Chicago Press, 1992).

11. The ultimate moral for aesthetics is that an aura in art is never something that happens purely because of a mechanical apparatus. The film apparatus (camera, shot, editing, screening) is for all intents and purposes the same as that of TV, whereas the aesthetic results of these media differ dramatically and fundamentally oppose each other. An aura is the effect of a total art form: from the apparatus of reproducibility in film's case to its screenplay, story, beginning-middle-end, composition of shots, placement of the star, etc. etc. . . . . If *Days of Our Lives* develops an aura over time, or *Oprah*, it will be because it becomes a cult environment, a kind of theater. Perhaps Diana carries the aura of stardom because royalty is also a thing (largely) of the past. She wears its historical mantle.

8. Diana Haunted and Hunted on TV

1. Andrew Morton, *Diana: Her True Story* (New York: Pocket, 1998), pp. 353–355.

9. Star Aura in Consumer Society (and Other Fatalities)

1. For a superb introduction to this desultory state of affairs, mostly concerned with Picasso's celebrity, see John Berger, *The Success and Failure of Picasso* (New York: Pantheon, 1989).
2. Richard Stolley, quoted in Joshua Gamson, *Claims to Fame: Celebrity in Contemporary America* (Berkeley: University of California Press, 1994), p. 43.

Index